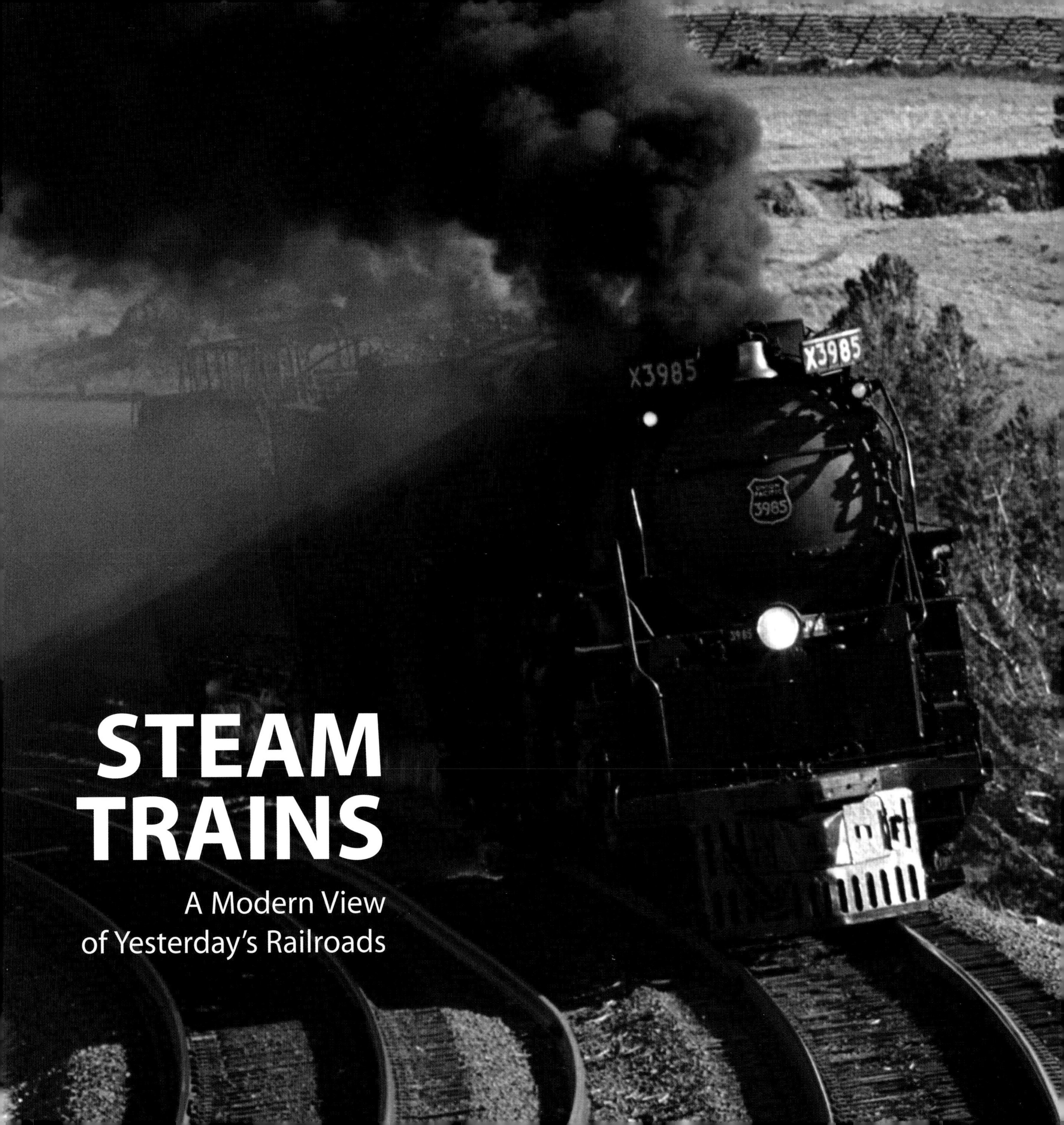

STEAM TRAINS

A Modern View
of Yesterday's Railroads

STEAM TRAINS

A Modern View of Yesterday's Railroads

James P. Bell

Voyageur Press

Design and Repro by Philip Clucas MSIAD
Edited by Mariam Pourshoushtari

This edition published in 2006 by Voyageur Press, an imprint of MBI Publishing Company, Galtier Plaza, Suite 200, 380 Jackson Street, St. Paul, MN 55101-3885 USA

MBI Publishing Company titles are also available at discounts in bulk quantity for industrial or sales-promotional use. For details write to Special Sales Manager at MBI Publishing Company, Galtier Plaza, Suite 200, 380 Jackson Street, St. Paul, MN 55101-3885 USA.

ISBN-13: 978-0-7603-2267-3
ISBN-10: 0-7603-2267-8

Printed in China

Right: UP No. 3985 has just pulled into the yards in Van Buren, Arkansas, on February 6, 2004, on the return trip to Cheyenne, Wyoming, after traveling to the Super Bowl in Houston, Texas.

Contents

Introduction

LEFT: Union Pacific Challenger 4-6-6-4, No. 3985, heads south into Kansas pulling the Super Bowl Special to Houston, Texas, on January 11, 2004.

Steam railroading in the twenty-first century remains as America's largest mobile art form. A steam locomotive at 70 miles per hour speeding across the landscape is poetry in motion. So many parts of a steam engine are exposed for the world to behold. The panting of air pumps and the sound of exhausts make a steam locomotive seem alive like no other modern method of transport.

Railroads established the patterns of development across North America. Place names along the lines recall the glory of a vigorous nation reaching west to pursue its manifest destiny. Can there be anything more magical than hearing the names of towns and cities along a transcontinental mainline and imagining them stretched out on shining twin rails like a string of pearls? The names of Dodge City, Trinidad, Raton, Albuquerque, Flagstaff, San Bernardino, Pasadena, and Los Angeles sing a romantic song of the Southwest.

Some people lament the fact that they were born too late to have witnessed the spectacle of mid–twentieth century steam railroading. Railroads affected the lives of so many people in so many ways across North America.

I can barely remember the sounds of steam switchers in my hometown of Fort Smith, Arkansas, when I was four or five years old. But that smell of coal smoke remained in my mind until I was acquainted once again with steam railroading in the 1970s.

Even though I missed the heyday of steam, I feel fortunate to be alive to see remnants of it today. Modern cameras, film, and the digital revolution have given us ways to capture an image of steam railroading impossible for those who came before us. A pantheon of legendary rail photographers placed the bar high for those who follow. Railroad photographers like O. Winston Link, Dick Steinheimer, David Plowden, Don Ball Jr., and Phil Hastings are just a few who set the standards for creative railroad photography.

But I wouldn't go back to the 1930s or 1940s even if I could. Even to witness the *Chief* climbing Raton Pass in 1936, I wouldn't. Those were tough times. Consider the state of medical care alone in those days. Polio confined thousands of young people to lives of disability. Children died of pneumonias that are easily treated today with outpatient antibiotics. Men and women died of heart attacks in their 40s and 50s with no offering of effective treatment. The wonders of CT and MRI scanners in medicine were the dreams of science fiction writers.

No, give me the present and the glories of steam railroading that persist into modern times. I've experienced not only a Union Pacific Challenger climbing Sherman Hill and a snowplow clearing the Cumbres Pass narrow gauge, but I've also witnessed men walking on the moon, streamlined trains in Europe traveling 120 miles per hour, and the miracle work of a neonatal intensive care unit—where my grandson received treatment that allowed him to survive a premature birth. I look forward to taking him to Cheyenne someday in the future to see 4-6-6-4s climbing mountain passes.

May there always be people in this country who can't forget that sound of a steam whistle in the night, the smell of hot grease, steam and coal smoke in an active railyard, and the feel of a giant engine vibrating the ground under their feet at trackside. May there always be people in the United States and Canada who can lovingly tend the great iron beasts from the past that provide us with these nations' largest mobile art form.

1 Spokane, Portland & Seattle No. 700 Returns to the "Last Best Place": Steam in Montana

"The Last Best Place" is how Montana editors described the Big Sky state in an anthology of literature celebrating their state's centennial in 1989. For those enamored with big-time railroading under the broad canvas of prairies and rugged mountains, Montana once offered three main theaters of transcontinental action. The Great Northern Railway crossed the state along the northern tier of counties called the High Line. The Milwaukee Road (officially the Chicago, Milwaukee, St. Paul & Pacific Railroad) once brought railfans and photographers to the steppes of central Montana to witness mountain railroading with Little Joes and electric box cabs. The Northern Pacific (NP) served the southern part of the state along the old Lewis and Clark Trail beside the Yellowstone River.

Conditions had changed by the dawn of the 1980s. The Great Northern and Northern Pacific had already morphed into the Burlington Northern (BN) Railroad. The Milwaukee Road disappeared altogether by 1985 due to bankruptcy. Architectural artifacts of this once vibrant transcontinental road vanished annually.

At one time in 1979, the state still supported two east-to-west Amtrak passenger trains, the *Empire Builder* across the High Line and the *North Coast Limited* from Billings, Montana, to Sandpoint, Idaho. The *North Coast Limited* did not survive into the 1980s due to budget cuts.

Steam railroading had been absent from Montana since the 1950s, but this was about to change with a spin-off shortline from the BN.

Right: SP&S No. 700 leaves Bozeman, Montana, under a veil of steam and smoke on the morning of October 14, 2002.

Left: Spokane, Portland & Seattle (SP&S) No. 700 rounds a bend south of Helena, Montana, en route to Bozeman pulling the Montana by Steam special on October 13, 2002.

700
700
700
WRONG
WAY

Citizens of communities along the old Northern Pacific probably feared the worst when the BN proposed giving up its southern route to a shortline operator. Residents of Livingston, Montana, were especially uneasy when their massive locomotive shops appeared to be headed toward extinction.

In spite of these worries, Montana RaiLink (MRL) came into existence in 1987 and emerged as one of the more vigorous Class II operations in the United States. Because of MRL management's openness to private operation of excursion passenger trains, steam finally returned to the "Main Street of the Northwest" in 2002.

Marcia Pilgeram, a Montana native from Trident, grew up close to the old NP. "I rode trains all the time when I was a kid," she said. "When I worked for Montana RaiLink, I used to organize special business trains."

A consortium of investors, formed in the mid-1990s, became Montana Rockies by Rail Tour Group. Marcia Pilgeram left MRL to become president and CEO of the rail passenger company based in Sandpoint, Idaho. With a streamliner consist and motive power supplied by MRL, the company offered first-class excursions from Sandpoint to Livingston. On special occasions the train traveled all the way to Billings, Montana.

In the year 2000, Pilgeram's chief mechanical officer, Ken Keeler, suggested that the company approach the Pacific Railroad Preservation Association (PRPA) about using Spokane, Portland & Seattle (SP&S) No. 700 for a steam trip. All parties involved—the BN, MRL, and the PRPA—thought the trip was a good idea.

LEFT: *SP&S No. 700 storms up Bozeman Pass in southern Montana early on the morning of October 14, 2002. The passenger special is en route to Billings, Montana under a cloudless canopy in the Big Sky state.*

ABOVE: *SP&S No. 700 disappears into a cloud of steam and smoke at a grade crossing on the western climb up Bozeman Pass in Montana on the morning of October 14, 2002.*

When Montana Rockies by Rail considered using a steam engine for a special tour, it "originally wanted to run during the winter months," according to Pilgeram. "But concerns about a breakdown out in the mountains and what you would do with a trainload of passengers in that situation lead us to split the difference and settle for October."

Locomotive No. 700 traveled from her home base in Portland, Oregon, to Sandpoint, Idaho, and the *Montana Daylight* by Steam tour pulled out eastbound the morning of October 12, 2002. About 200 passengers were onboard the train at any one time. By trip's end, some 660 people had ridden the steam special as the train traveled over the Continental Divide, down the Yellowstone River, to Billings, and back again.

Noted rail authors Karl Zimmerman and Al Runtes joined this historic train for lectures along the way. Bill and Jan Taylor, who were Missoula residents, authors, and Northern Pacific history experts, served as tour guides and offered their knowledge to the passengers who came from all over the country for this once in a lifetime event.

This steam excursion over the Yellowstone Park Line, as the former NP line was called, reminded passengers and observers of the railroad's close association with the country's first national park in 1872. People along the line turned out in crowds to watch an event that had been absent from their hometown tracks for 50 years.

RIGHT: SP&S steam locomotive No. 700, a 4-8-4 Northern type has exited the tunnel at Mullan Pass and is drifting downgrade over the graceful, curved bridge called Skyline Trestle on October 13, 2002.

700
700

TESLOW
INC.
GRAIN
HAY

Montana Rockies by Rail picked the right week for a steam train excursion. A storm front blew through the night before departure and deposited fresh snow on the mountains. Then the skies cleared to a deep blue, and visibility became the usual 75 miles or greater. The aspens, cottonwoods, and larch provided a golden aura in the autumn sunshine.

No. 700 pulled a flawless trip both east- and westbound. Whether it was steaming over such landmarks as the elegant curved bridge called Skyline Trestle, storming up Bozeman Pass, or pausing at preserved depots in Helena, Bozeman, Livingston, and Billings, No. 700 became the star in creating classic images from the past.

At the end of the rail excursion, Pilgeram called the trip "perfect." She recalled, "I can't tell you how thrilling it was to hear the stories of people riding that train. People would show up at the stations and cry when they saw the engine, as they remembered riding trains along this line long ago." Steam returned for a season to the rails of the Main Street of the Northwest through Montana.

BELOW LEFT: SP&S No. 700 prepares to leave Missoula, Montana, on the morning of October 13, 2002.

BELOW RIGHT: A curious resident of East Helena, Montana, watches for SP&S No. 700 as she leaves the capital city of the Big Sky state on October 13, 2002.

LEFT: SP&S No. 700 picks up speed leaving Livingston, Montana, at midday on October 14, 2002. The Montana by Steam special stopped in Livingston to allow passengers to tour the museum in the restored Northern Pacific depot.

2 An Arkansas Shortline Survives: Celebrations on the Dardanelle & Russellville

On a Sunday, August 21, 1994, the Dardanelle & Russellville (D&R) Railroad celebrated its 111th birthday by offering passenger train rides over the entire railroad system. The special train made four or five roundtrips in one afternoon. This spunky railroad is not only the oldest surviving shortline in Arkansas, but also perhaps the shortest at 4.8 miles in length.

A genuine smoke-belching, fire-breathing steam locomotive powered the passenger train, while the railroad's diesel sat idle. The shortline obtained locomotive No. 4 from the Reader Railroad and subsequently lettered the engine for the D&R.

"We leased No. 4 from the Reader in 1993," said owner of the railroad, Bill Robbins, as he recalled how a steam engine returned to the D&R. "The locomotive did not ever run at Reader," said Robbins. "We completely overhauled her here in our shop."

The birthday-party passenger train trundled out to the Union Pacific (UP) Railroad's mainline connection with the D&R in downtown Russellville. The return trip brought riders back to the D&R home offices on the north bank of the Arkansas River, across from the town of Dardanelle, Arkansas, the other part of the railroad's namesake. Private owners of rail speeder cars were invited to join the party, and they traveled along behind the train.

LEFT: Railfans come out to the offices and shops of the Dardanelle & Russellville Railroad to help celebrate the 111th anniversary of the shortline railroad. Locomotive No. 4 of the D&R is in steam and giving rides to downtown Russellville.

RIGHT: Dardanelle & Russellville steam locomotive No. 4 steams past one of the vintage concrete posts that supports the railroad crossing signs in Russellville, Arkansas, in August 1994.

D&R
4
ROAD

D&R
4

LEFT: Railfans gather around the Dardanelle & Russellville steam locomotive No. 4 on a Sunday afternoon in August 1994 as the railroad celebrates its 111th anniversary.

Baldwin Locomotive Company built the D&R No. 4 in 1913 for the Laurel River Lumber Company of West Virginia. For unknown reasons, the steam engine—a 2-6-2 Prairie type—was returned to Baldwin in 1914. W. T. Carter and Brother Lumber Company in Camden, Texas, purchased the locomotive, and she moved to the piney woods of east Texas in 1914.

For the next 60 years the locomotive hauled logs from the woods to lumber mills. In 1973, Richard Grigsby of Malvern, Arkansas, brought four of the lumber company's steam engines up from Camden to Arkansas. Two were for use on his newly purchased Reader Railroad, and two were for steam enthusiast Robert Dortch of Scott, Arkansas. Bill Robbins assisted in this move and became acquainted with No. 4 before he owned the railroad to which the locomotive would one day come.

The steamers that Grigsby brought up from Texas were No. 1 and No. 201. A cabbage stack wood burner, No. 1 ended up on Dortch's Scott and Bearskin Lake Railroad and eventually was moved to the Eureka Springs & North Arkansas (ES&NA) Railroad. The second locomotive, No. 201, also came to the ES&NA where it remains today. Grigsby lettered No. 2 and No. 4 for his Reader Railroad.

BELOW: D&R No. 4 originally came to the shortline from the Reader in southern Arkansas. The locomotive is found between trips under steam in August 1994 at the offices of the Dardanelle & Russellville Railroad, on the north bank of the Arkansas River.

RIGHT: Dardanelle & Russellville No. 4 is under steam between trips from the shortline's offices and Russellville, Arkansas.

ABOVE: Dardanelle & Russellville No. 4 steams back into the yard area of the railroad on the north banks of the Arkansas River across from Dardanelle, Arkansas, in August 1994.

Coal became king in the land between Dardanelle and Russellville, Arkansas, at the turn of the century. A railroad was needed to haul the semi-anthracite coal from the mines and haul miners to the coal. The Dardanelle & Russellville Railway first formed in 1893 and reorganized under its current name in 1900. The D&R filled the need for coal-related transportation for the next 50 years.

Bernice Anthracite Coal Basin and the Pittsburg & Midway Coal Mining Company provided traffic for the D&R in the early years of the railroad's existence. The D&R also owned the pontoon bridge spanning the unruly Arkansas River between the Russellville side on the north and Dardanelle on the south bank. This bridge carried people, wagons, and motor vehicles only. Workers came across this bridge for the mines, and they rode the trains of the D&R to work.

The advent of automobiles and the public clamor for better roads sounded a warning for the D&R's passenger traffic. In 1920, the shortline carried 99,902 passengers along its 4.8-mile-long route. In 1927, the Arkansas Highway Department completed a new steel bridge over the Arkansas River that replaced the D&R's pontoon toll bridge. D&R passenger traffic dropped to 13,000 that year. The coal mines shut down in the 1950s, and the last passenger train operated over the railroad in 1960. Today, the D&R continues hauling freight to the Union Pacific interchange in Russellville.

4
4
4

4
4
4

Bill Robbins, owner and operator of the D&R, is quick to point out that of the five recorded steam locomotives ever owned by the Dardanelle & Russellville Railroad, three still exist.

The surviving D&R locomotives include No. 8, a 4-4-0 bought from the St. Louis, Iron Mountain & Southern (SLIMS) Railway in 1900. This locomotive moved from Russellville to Southwest City, Missouri, for a time to star in 20th Century Fox's movie, *Jesse James*. The movie also starred Henry Fonda and Tyrone Power. No. 8 is today at the Nevada State Railroad Museum.

The Civitans Club of Shreveport, Louisiana, purchased D&R steam locomotive No. 10 in 1957. The engine still exists, although it is dismantled and stored. No. 9 remains at the Mid-Continent Railway Museum, which bought her in 1963.

Bill Robbins started his professional career as a pilot and aviation mechanic. He sold Piper airplanes all over the state of Arkansas in the 1960s. Robbins left aviation in the mid-1980s and settled into raising breeder hens for a poultry business. But the challenge was not present. "I had always been fascinated by the shortline railroad business," he recalled as he reflected on how he came to buy the D&R. After purchasing the D&R in 1988, he also bought another Arkansas shortline, the Ouachita Valley.

When Robbins was a young man in 1959, he had a chance to buy D&R No. 9. Along with fellow Fort Smith rail enthusiasts, Louis Marre and Gordon Mott, Robbins inspected No. 9 in Dardanelle when it was for sale. Unfortunately, the men did not have a railroad on which to run the locomotive at the time. "I look back at that chance and wonder what if?" said Robbins upon reflection of passing on the chance to buy that locomotive.

LEFT: Dardanelle and Russellville locomotive No. 4 is under steam and waiting for passengers to board for the next run from the shortline's offices at North Dardanelle.

3 Absorption of the Missouri Pacific into the Union Pacific Brings a Return of Steam to Arkansas

The 1980s resulted in several significant mergers by Class I railroads in America. The absorption of conservative, dark-blue Missouri Pacific (MP) Railroad into the expansive, Armour Yellow Union Pacific (UP) System gave railfans in the heartland a chance to experience mainline steam railroading once again. The MP merger into the UP System occurred in 1982. In March 1984, UP No. 8444 left her home in Cheyenne and steamed south to the New Orleans World's Fair, officially called the 1984 Louisiana World Exposition. UP planned to take its steam engine to a broader audience in the newly acquired MP territory.

Across Arkansas, the UP *World's Fair Special* followed the route of the old *Texas Eagle* and *Sunshine Special* running south from St. Louis to Little Rock, Arkansas. On March 13, 1984, the *World's Fair Daylight* passed through such historic MP towns as Newport, Arkansas, where the train crossed the White River, one of the major drainages of the Ozark Mountains. At Kensett and Searcy, Arkansas, the steam train passed places where the MP once shipped tons of fresh Arkansas strawberries to the country in years past.

Entering Little Rock, the UP passenger train crossed the Arkansas River and pulled onto the tracks in front of Union Station. The steam engine and her coaches were moved over to the UP shops in North Little Rock for the night before the next day's run into northern Louisiana.

RIGHT: Camera crews provide the light for an early evening portrait of Union Pacific No. 8444 in Little Rock, Arkansas, on March 13, 1984. The train is en route to be displayed at the 1984 Louisiana World Exposition in New Orleans.

BELOW: Union Pacific's Northern type locomotive No. 8444 leads the World's Fair Special past grain silos in northeastern Arkansas on March 13, 1984.

X8444
X8444
UNION
PACIFIC
8444

8444
8444
UP
FEF-3-80-25/32-266

ABOVE: *A member of the engine crew steps down from the massive tender of UP No. 8444 in North Little Rock, Arkansas, on March 14, 1984. The locomotive from Cheyenne, Wyoming, is prepared for another day on the road to the 1984 New Orleans World's Fair.*

LEFT: *Engine crews for Union Pacific No. 8444 and the* World's Fair Special *discuss preparations for the day's trip from North Little Rock, Arkansas, on March 14, 1984.*

Southern Pacific's (SP) No. 4449 and its *World's Fair Daylight* joined No. 8444 in New Orleans. The SP train spent only a week at the fair and then steamed back to Portland, Oregon.

From March to November, millions of visitors toured the 1984 Louisiana World Exposition on the banks of the Mississippi River. Besides the two steam trains visiting the fair, the exposition hosted the U. S. space shuttle *Enterprise*. The spacecraft flew piggyback on top of a Boeing 747 and landed at an airport near Mobile Bay, Alabama. The shuttle was loaded by crane onto a barge and brought along the inter-coastal waterway to the mouth of the Mississippi River. The barge and spacecraft then traveled up river to New Orleans where the shuttle was unloaded and hoisted to its site at the fair.

Other methods of transportation to the fair included paddlewheel riverboats and a monorail. The New Orleans World's Fair proved instrumental in spurring redevelopment along the central business district of the city.

When UP No. 8444 left Louisiana and returned to Wyoming, thousands of Americans had been reintroduced to the thrill of steam trains on mainline railroads. The trip to the south inspired the renovation of other steam locomotives around the country.

X8444
X8444
UNION
PACIFIC
8444

ABOVE: The World's Fair Special on the Union Pacific Railroad has pulled into Little Rock's Union Station on the evening of March 13, 1984. The train is en route to the 1984 New Orleans World's Fair.

LEFT: UP No. 8444 storms across the White River bridge near Newport, Arkansas, on the afternoon of March 13, 1984. The locomotive is pulling the World's Fair Special. It was on display from May until November 1984 in the 1984 Louisiana World Exposition in New Orleans.

Visitors to the UP's exhibit at the World's Fair learned that this 4-8-4 Northern never left active service. In 1960, as UP completed the replacement of steam with diesels, No. 8444 found reprieve by a season of use as a stationary steam boiler for railroad facilities in Cheyenne. The locomotive even had its original number of 844 changed to 8444 to avoid conflict with a diesel numbered 844.

No. 8444 continued in service as railfan groups chartered the locomotive and the UP streamlined coaches for excursions in the Denver–Cheyenne area. The annual Denver Post Frontier Days Special remained a favorite. One of the two active UP steam locomotives, either No. 8444 or UP No. 3985, usually shared the honor of pulling this train from Denver to Cheyenne and back on the opening Saturday of the Frontier Days Rodeo each July.

After the Arkansas visit of No. 8444 in 1984, both UP steam engines regularly roamed the expanded UP system each year. Visits to National Railway Historical Society conventions around the country gave the UP trains an opportunity to open the world of steam railroading to a new generation of people.

BELOW: *The crews of UP No. 8444 have climbed aboard the tender of the 4-8-4 Northern from Cheyenne, Wyoming, as the* World's Fair Special *is service in Newport, Arkansas, on the afternoon of March 13, 1984.*

Steve Lee shepherded the move of UP's multi-million-dollar steam train to New Orleans in 1984. He had been with the UP for only two years when he assumed his job as manager of train operating practice. "I didn't chose the job, I was assigned," said Lee. He started his life in railroading the traditional way, by taking a job with the Illinois Central in the early 1970s in his home state of Kentucky. He then moved further west by joining the Rock Island in the Kansas City area as a dispatcher. He left the ailing Rock for the UP three weeks before bankruptcy closed the Rock Island for good in 1980.

Although Lee didn't come to railroading as a steam enthusiast, he finds maintaining and operating the UP's vintage equipment a challenge. He figures the steam locomotive operating at the time, either No. 8444 or No. 3985, averages between 5,000 and 10,000 miles each year.

When asked why he continued in his job of steam railroading, Lee replied, "I must like the job I do, or I'm nuts." He continued by saying, "It has a lot of variety. You don't just do the same thing every day. You see a lot more of the railroad [when you run the steam engine].

"I added them up recently and we've been to all of the western states except for Montana, Washington, and Arizona," said Lee, as he thought about his unique job as a steam railroader in the new millennium.

Lee noted that the steam program is in it 46th year of operation. He has worked with steam masters Lynn Nystrom and Ed Dickens over the years. "The commitment from the UP is stronger than ever," commented Lee, when asked about the continuation of steam on the expanded UP system. "I'm looking forward to leaving the operation in good hands when I retire," he said.

On a final note, UP No. 8444 received her previous number of 844 when the diesel bearing those three digits was retired in 1989.

Above Left: As evening light envelopes the train of the World's Fair Special *in Little Rock, Arkansas, passengers leave the warm confines of a Union Pacific dome car to travel to hotels in Little Rock, Arkansas, on March 13, 1984.*

Above Right: Union Pacific No. 8444's distinctive smoke deflectors sported a large white sign carrying the name of the train, the World's Fair Special. *The train is prepared for departure from North Little Rock, Arkansas, on the morning of March 14, 1984.*

201
EUREKA SPRINGS & NORTH ARKA

4 A Railroad War in the Ozarks: A Tale of Two Shortlines with Claims on Eureka Springs

Bob Dortch Jr. farmed for much of his life in Scott, Arkansas, on the family's historic plantation outside of Little Rock. He always had an interest in heavy equipment due to the nature of his work. Dortch became fascinated with steam tractors and tinkered with several on his farm. In 1973 he assisted Richard Grigsby of Malvern, Arkansas, in bringing three steam locomotives to Scott from Camden, Texas. On newly laid track, Bob began running a tourist train out to Bearskin Lake about 2 miles away.

Two of Dortch's three locomotives on the Scott and Bearskin Lake Railroad were No. 1, a cabbage stack wood burner, and No. 2, an oil-burning sister engine to No. 1. Both were 2-6-0s built by Baldwin for the W. T. Carter and Brother Lumber Company in 1907. The third engine was another 2-6-0 type, No. 201, built for Panama Canal construction but later brought to the Moscow, Camden & St. Augustine Railroad in Camden.

Dortch had a plan: he would move his three steam locomotives and an assortment of coaches from his plantation on the Arkansas River to Eureka Springs, Arkansas. He wanted to replant his tourist railroad and his family to the popular mountain town referred to as "Little Switzerland of the Ozarks."

Left: Eureka Springs & North Arkansas No. 201, 2-6-0, leads one of the daily excursion trains back into Eureka Springs, Arkansas. The locomotive came out of the logging woods of east Texas from the Moscow, Camden & San Augustine Railroad.

Right: ES&NA No. 201 is seen between runs at the Eureka Springs depot in northern Arkansas. American Locomotive Company built the 2-6-0 in 1922.

RIGHT: Eureka Springs Railroad 0-4-0 T No. 3 rolls across the entry to the White River Bridge in September 1983.

ABOVE: ES&NA No. 201 steams proudly to a stop in front of the platform at the restored stone depot in Eureka Springs, Arkansas, in the early 1980s.

When the Dortch family first moved to Eureka Springs, it started a modern day railroad war, according to Craig Marheineke, a former Eastern Shore Railroad (ESRR) engineer. Reat Younger of Springfield, Missouri, already operated a tourist steam train at the small community of Beaver, Arkansas, on the upper reaches of Table Rock Lake in the White River valley.

Since 1978, Younger and his brother Dreat operated a 0-4-0T switcher pulling a string of cabooses and a coach from the town of Beaver, across the river, and to a high rock cut called the "narrows." Along the right of way of the former Missouri & North Arkansas (M&NA) Railroad, 7 miles of deep canyons, numerous trestles, and a tunnel separated the narrows from Eureka Springs. There was only one problem in getting into the town of Eureka Springs by rail from the narrows. The tracks had been removed when the M&NA was abandoned.

Bob Dortch came to Eureka Springs in 1980, obtained a lease on the ornate limestone depot at the bottom of the mountain in the town, and began laying the tracks of his newly formed Eureka Springs & North Arkansas (ES&NA) Railroad toward Beaver and the Eureka Springs Railroad. For a time Dortch tried to convince Reat Younger to combine their two tourist lines into one. Younger and Dortch were said to be like oil and water. A merger never happened.

3
3

About 2 miles out of Eureka Springs was the wye at the junction to the old M&NA mainline. This was where the new ES&NA tracks stopped. A contested lease over rights to this wye ended in a court battle between Dortch and Reat Younger. Eventually Dortch won the case, but Craig Marheineke remembered it as being based on "possession is nine-tenths of the law." Both railroads operated simultaneously for a few years.

Marheineke came to the Eureka Springs Railroad right out of high school from his hometown of Rogers, Arkansas, in 1981. "I was always interested in trains, but the Frisco was my railroad. I didn't even know the North Arkansas existed," said Marheineke. He trained first as fireman, then engineer, and finally conductor. He obtained his Arkansas boiler license in 1982 and worked as one of two full-time engineers through the seasons of 1984 and 1985.

Marheineke came to know Reat and Dreat Younger fairly well through his three years of employment on the railroad. The Youngers also had two other brothers, Cleat and Gleat (who went by his other name of Jerry), who were not involved in the railroad business. But the Younger family definitely had railroad connections from birth.

Reat was born in the Higdon, Arkansas, depot on the old M&NA, the same line their Eureka Springs Railroad used. "The Younger brothers' father was a traveling agent for the M&NA," said Marheineke. "Dreat worked as a news butcher, apprentice fireman, and then fireman on the line."

Right: Eureka Springs Railroad No. 3 with a diamond stack crosses the steel trestle over the White River on the upper end of Table Rock Lake at Beaver, Arkansas, in September 1983.

ABOVE: ES&NA No. 201 stops in front of the water tank in front of the Eureka Springs depot.

RIGHT: The No. 214 wooden caboose of the ES&NA came from the Louisiana and Northwest Railroad. The caboose was originally built for the Cotton Belt Railroad in the early 1900s.

Above: *ES&NA No. 201 is between runs at the railroad line that formerly belonged to the Missouri & North Arkansas Railroad. The restored stone depot is headquarters for the tourist line in Eureka Springs, Arkansas, also called the "Little Switzerland" of the Ozarks.*

Left: *Between trips, ES&NA No. 201 yields a detail inspection of a driving wheel and side rods in the Eureka Springs, Arkansas, depot.*

As of 2006 the Eureka Springs & North Arkansas Railroad is the survivor in the railroad wars in the Ozarks. "Our business just tanked when the ES&NA opened up," said Marheineke. After the 1985 season, the Younger brothers operation at Beaver ceased.

The little 0-4-0T, No. 3, at Beaver was originally brought in by truck from the Long Bell Lumber Company in Joplin, Missouri. Eventually No. 3 left the White River valley in the same way she came, on a low-boy trailer truck to a new owner in Springfield, Missouri

David Dortch and his brother, Robert III, continue to run the ES&NA and make their home in this quaint Victorian town in the Ozarks. The brothers' father, Bob, who brought them to Eureka Springs as young men just out of school, passed away in 2002. David just recently returned to Eureka Springs after being in California for two years while restoring private rail coaches for Amtrak use.

All three steam locomotives at the ES&NA are idle and need mechanical work. The excursion trains and the popular dinner train on the line are now handled by a diesel engine. Will steam ever return to the ES&NA? "Perhaps," said David, "when another Bob Dortch [Jr.] comes along."

ABOVE: Frisco No. 1522, a Mountain type locomotive gives residents of Gainesville, Texas, a taste of steam railroading on June 7, 2001, as the Burlington Northern Santa Fe Employee Appreciation Special passes through the center of town.

RIGHT: The streamlined train of the BNSF Employee Appreciation Special rumbles through the central Oklahoma countryside on June 8, 2001.

5 Frisco No. 1522 on the Route of the Heartland Flyer in Oklahoma

In this age of the internet and electronic commerce, where even the term "information superhighway" seems slightly antiquated, it is easy to forget that another industry once shaped and molded the geographical and cultural map of America as it exists today.

Railroads in the past held as much power as computer and internet companies do today. Just as the digital revolution is changing lives of people around the world, at one time railroads determined how and where people lived when the twentieth century dawned. As the transcontinental lines neared completion, branch lines fanned out into the hinterland. These feeder lines brought small towns their merchandise and mail and took the farmer's milk and produce to market.

Some villages in North America packed up their entire communities and moved to meet an arriving rail line that bypassed their original town site. Railroads often determined whether small towns developed and thrived or passed into oblivion.

In the 1970s and 1980s Amtrak and VIA began serving the needs of passenger rail transportation in North America. In the cities where these government-subsidized rail services remained, the depots and stations they used survived and thrived. When passenger services ceased, the razing of many community railroad stations soon followed.

At its inception in 1971, Amtrak provided service through Oklahoma on a north–south railroad line. After eight years, Amtrak abandoned this old Santa Fe route through Oklahoma's Arbuckle Mountains. As expected, historic depots along this route were neglected and left to decline.

Right: Frisco No. 1522 picks up speed at Red Rock, Oklahoma, on June 8, 2001.

With grassroots support and state funding, Amtrak brought passenger service back to the Sooner State in 1999. A new train, the *Heartland Flyer*, retraced the old route of the Santa Fe Railroad's *Texas Chief* and *Kansas Cityan* between Oklahoma City and Fort Worth, Texas. In collaboration with Amtrak, communities resurrected former abandoned passenger stations. These depots became centers of pride and commerce for the communities they served.

Restored depots greeted the Amtrak traveler at Oklahoma City, Norman, Paul's Valley, Ardmore, and just over the line into Texas at Gainesville. A new station has even been built at Purcell. Each of these depots now serves as a magnet and focus for community activity. People still turn out each day to watch the arrival of a train.

Public support for Oklahoma's Amtrak *Heartland Flyer* has been outstanding, with ridership approaching 70,000 passengers per year. People come from a wide corridor across the central part of the state to ride the train. Travelers use the *Flyer* to connect to trains at Fort Worth bound for the rest of country. The *Heartland Flyer* is important to the citizens of Oklahoma and Texas.

The significance of passenger railroad stations in improving the life within small rural communities became clear when Frisco steam locomotive No. 1522 traveled the route of the *Heartland Flyer* carrying a Burlington Northern Santa Fe (BNSF) Railroad *Employee Appreciation Special* in June 2001.

RIGHT: On June 7, 2001, the Burlington Northern Santa Fe's Employee Appreciation Special *pauses for passengers at the restored Santa Fe depot in Ardmore, Oklahoma. The Ardmore Main Street Project now shares the offices with the Amtrak passenger station.*

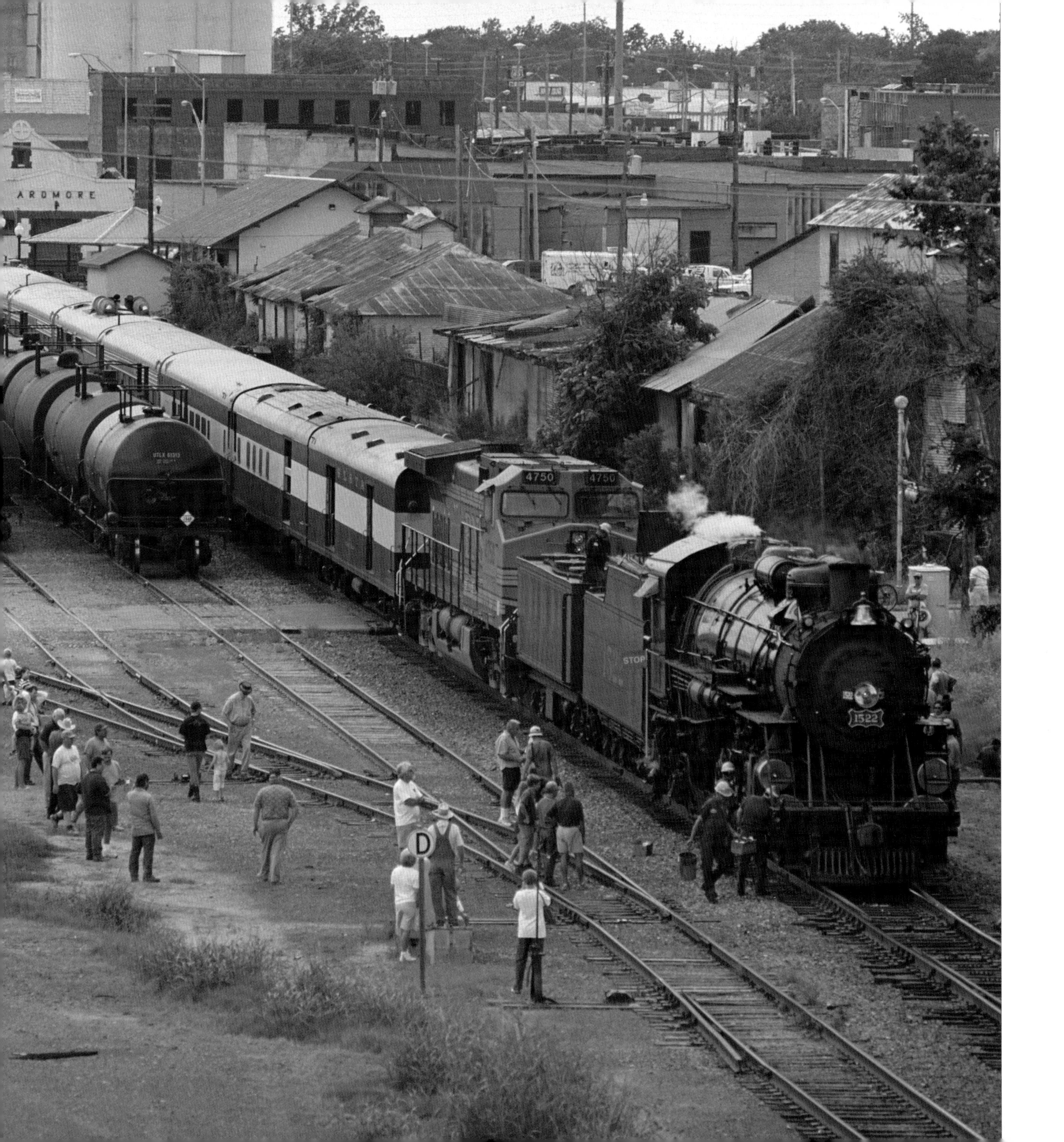
ARDMORE
4750
4750
1522
D

ABOVE: The renovated Santa Fe depot in Oklahoma City anchors the entrance to Bricktown. The BNSF Employee Appreciation Special *lead by Frisco 1522 spent the night June 7, 2001, in Oklahoma's capital city after a day of travel from Fort Worth, Texas.*

RIGHT: Frisco No. 1522 and the BNSF Employee Appreciation Special *steams past the new depot in Purcell, Oklahoma. When the depot is not welcoming steam specials, the facility serves as the Amtrak depot for patrons of the Heartland Flyer.*

When Frisco No.1522 steamed back north across the heart of Oklahoma pulling the BNSF *Employee Appreciation Special*, the differences between south-central Oklahoma and north-central Oklahoma became apparent. South of Oklahoma City, current Amtrak depots served as vibrant community centers. North of Oklahoma City where rail passenger service was gone, the depots of the old Santa Fe Railroad seemed neglected and forgotten. The contrast showed how active passenger station brought life to the center of the town it served.

On Thursday, June 7, 2001, Frisco No. 1522 departed Haslet Yard in Fort Worth, Texas, for the gallop north toward home in St. Louis. At Gainesville, Texas, a crowd gathered with folding chairs on the grassy lawn in front of the freshly restored Santa Fe depot and waited for the steam special to appear. Schools brought their students to the depot to witness a rare event in their town's history. Heart rates rose when a headlight appeared around the bend south of the depot. But it was only Amtrak, *Heartland Flyer*, running ahead of the steam special. The *Flyer* stopped briefly at its last northbound Texas stop to board passengers bound for Oklahoma. The passengers on board the *Flyer* probably had no clue that a steam train was right on their tail.

Soon Frisco No. 1522 and her *Employee Appreciation Special* roared around the bend in the tracks to the south. She steamed past the Gainesville station and did not disappoint those who had gathered to watch her passing. This Mountain type engine built by Baldwin in 1926 was especially known for her distinctive sharp stack exhaust sound. In a billow of smoke and steam, No. 1522 made a grand departure from Gainesville

1522

1522
1522

Above: On June 8, 2001, the BNSF Employee Appreciation Special *pauses in front of the former Santa Fe depot at Perry, Oklahoma. The depot is not used at the time and windows and doors have been secured with plywood. Fortunately the depot has not been razed and waits for future redevelopment.*

Left: The BNSF Employee Appreciation Special *powered by Frisco No. 1522 runs at 50 miles per hour on straight track through the heart of Oklahoma on June 7, 2001. This would be one of the last major trips run by the Mountain type locomotive from the Museum of Transportation in St. Louis, Missouri. The steam engine was retired in 2002.*

Frisco No. 1522 picked up speed out of Gainesville. The train passed through Ardmore, Oklahoma, a thriving town built on oil and agriculture. The Main Street Association serving the city now occupies the former Santa Fe depot.

At Paul's Valley, a small wooden depot greeted rail passengers, while at the end of Main Street in Purcell, Oklahoma, a sparkling new brick depot served them. Norman, Oklahoma's former Santa Fe depot still stands in this city, which is home of the University of Oklahoma and its famous Sooner athletic teams.

In Oklahoma City, the Amtrak passenger is greeted by a scrubbed and renovated Santa Fe station built on a grand scale in the 1920s. The restored depot anchored the entry to the city's new Bricktown, a commercial development of vintage warehouses that are now upscale shops and restaurants.

After an overnight servicing stop in Oklahoma's capital, the steam train ran north out of Oklahoma City and remained on BNSF's former Santa Fe route. Depots in Guthrie, Ponca City, and Perry still stood, but the doors and windows were boarded and the stations appeared abandoned. The former Santa Fe depots stood as mute evidence to the premise that regular rail passenger service brought more to a community than just another mode of transportation. Railroad depots can still be used as community centers and serve as a central gathering place for citizens in small towns across the heartland of America.

6 Down Along the Cotton Belt

In the mid-1980s Pine Bluff, Arkansas, became a mecca worthy of the most devoted steam pilgrim. Cotton Belt Northern No. 819 was brought back to life and ran on frequent excursions across Arkansas and Texas. Pine Bluff, a city of 50,000 people, joined the list of American places—like Cheyenne, Wyoming, and Chama, New Mexico—where a North American steam locomotive could still be found working on home tracks.

The Cotton Belt built locomotive No. 819 in Pine Bluff in 1942. America was at war, and the railroad had been in bankruptcy since 1935. These were hard times. The War Production Board would not let the railroad buy new diesel locomotives, yet the Cotton Belt could build 10 4-8-4s modeled after Baldwin Northerns that it had purchased in the 1930s. These locomotives received numbers from 810 to 819. No. 819 became the last one built in the series. She first ran on February 8, 1943.

The Cotton Belt, officially called the St. Louis Southwestern Railway, became part of the Southern Pacific system through a merger in 1932. The railroad was eager to completely dieselize. After recovering from receivership after World War II, the Pine Bluff–built Northerns were soon shunted off to other places with several going to Southern Pacific's California commuter service. Most of the 4-8-4s would die by the scrapper's torch. No. 819 ended her journey in Pine Bluff's Oakland Park, protected by a shed built by Cotton Belt employees.

RIGHT: Cotton Belt No. 819 has arrived in Jonesboro, Arkansas, on June 12, 1990, on the way north to the National Railway Historical Society Convention in St. Louis. A detail shot from above shows pipes and bell clustered around the stack of the 4-8-4 Northern from Pine Bluff, Arkansas.

ABOVE: Cotton Belt No. 819 is caught at milepost 350 near Buena Vista on the Cotton Belt mainline on October 1990. The steam train from Pine Bluff is en route to Tyler, Texas.

RIGHT: Cotton Belt's restored Northern type No. 819 crosses the overpass above U.S. Highway 79 at Camden, Arkansas, and slows for a stop to board passengers. The train is en route to the Rose Festival in Tyler, Texas, on October 15, 1993.

The sleeping locomotive might have remained in the city park had it not been for a Southern Railway excursion in the early 1980s. Darrel Cason of Pine Bluff remembered, "I was riding behind the Savannah & Atlanta No. 750 on a trip from Memphis to Iuka, Mississippi, and I noticed a plaque on the tender which read 'This engine was restored and is owned by the Atlanta Chapter of the NRHS.'" Cason found Bill McCaskill—a Cotton Belt conductor—and Jim Bennett, both of whom were with him on that train, and said, "Why don't we do that with 819?"

On his return home, Cason called Jim Johnson, public relations officer with the Cotton Belt, and asked, "What do you think about forming the 819 Rail Historical Society?"

Cason and his friends found the community receptive to their idea. Bill Bailey, president of the Arkansas Railroad Club, had previously approached the city of Pine Bluff about the possibility of doing a cosmetic restoration of the locomotive in the park. The idea of restoring No. 819 to operational status met with the enthusiastic formation of the Cotton Belt Rail Historical Society.

Bailey and Cason became co-directors of the group called "Project 819." Other company employees soon joined the leadership of the endeavor. They eventually moved the steam engine from Oakland Park to the Cotton Belt erection shops in October 1983. Robert McClanahan and Joe McCullough, former Cotton Belt employees, provided their expertise throughout the project.

Left: On October 15, 1990, Cotton Belt No. 819, the "Pride of Pine Bluff," crosses Main Street in the city where she was built in 1942. The 4-8-4 is maintained at her birthplace in the shops of the old St. Louis Southwestern Railway, in Pine Bluff, Arkansas.

BELOW RIGHT: In Pine Bluff, Arkansas, Darrel Cason readies the Cotton Belt No. 819 for its journey to Tyler, Texas, on the morning of October 19, 1990.

BELOW: The old Cotton Belt Freight House in Brinkley, Arkansas, reminds people that the railroad deals in "Fast Freight." The building, pictured here in June 1990, was razed several years after this photograph was made.

"The locomotive was in great shape when she was put in the park," recalled Cason, referring to the 819s placement into Oakland Park. "We had less than $200 when we moved that engine." During the renovation, the group's talents as fundraisers were tested as much as their mechanical skills, as the renovation eventually cost $140,000.

Finally after nearly three years of hard work, No. 819 made her first trip to Fordyce, Arkansas, on April 25, 1986, for the Fordyce on the Cotton Belt, a fledgling festival celebrating the small community's ties to the railroad. The trip was a success and would become an annual event for several years.

A particular pleasure for passengers and rail photographers alike is to ride behind and photograph a steam engine on its native road. Railfans and the Cotton Belt group had this chance in June 1990 when No. 819 pulled a streamlined train from Pine Bluff to St. Louis for the National Railway Historical Society convention.

Above: "Fresh Fish Daily" is guaranteed by the sign for a fish market housed in a former automobile service station in Pine Bluff, Arkansas, on October 15, 1993. Cotton Belt No. 819 proudly steams through town en route to the Rose Festival in Tyler, Texas.

Cason said that money earned from the St. Louis excursion was used to repair the roof of the vast shop that protected the steam engine and the Historical Society's passenger consist. This shop and its growing collection of railroad equipment evolved into the Arkansas Railroad Museum. Many of the passenger coaches used in those early days of the museum were privately owned and leased to the historical society.

Another favorite run for the Historical Society for several years was a trip to Tyler, Texas, and that city's Rose Festival each October. "Tyler really treated us well," said Cason. "They enjoyed our coming down there." When she visited Tyler, No. 819 was always a star attraction for the city's annual event.

The railroad mergers and the liability insurance crisis of the 1990s adversely affected the Cotton Belt Rail Historical Society. In 1995, No. 819 could not run any excursions because Cotton Belt Railroad demanded liability insurance premiums that were unaffordable. In 1996, the Union Pacific Railroad absorbed the Southern Pacific and its subsidiary Cotton Belt. The Pine Bluff offices and freight depot were demolished, but the shops comprising the Arkansas Railroad Museum escaped flattening. These buildings continue to house No. 819 and an assortment of other equipment.

By the late 1990s, No. 819 needed routine maintenance repairs. With the increased difficulty of running on tracks now owned by the Union Pacific, the money and energy to keep the Cotton Belt Northern steaming became more difficult to obtain and justify.

For now, No. 819 is protected in the museum from the elements. The steam engine lies in wait with No. 336, a 2-6-0 built by Baldwin in 1909.

Will Cotton Belt No. 819 ever run again? No one knows for certain. There is only hope. Until that time, a cadre of dedicated volunteers lovingly tends the equipment at the Arkansas Railroad Museum and the Cotton Belt Rail Historical Society in Pine Bluff, waiting for the day when a homegrown 4-8-4 rumbles again across wooden trestles at 60 miles per hour in the delta swamps of eastern Arkansas.

RIGHT: A freight train takes the mainline near Big Sandy, Texas, as crowds leave after the passing of Cotton Belt No. 819 and her train en route to Tyler, Texas, on October 15, 1993.

261
261

7 Wayzata, Minnesota, Invites a Superstar: Milwaukee Road No. 261

Wayzata, Minnesota, knows how to throw a street party. On September 11–12, 1999, the citizens of this Twin Cities suburb closed off their main drag through town, organized a parade, and invited a superstar. The superstar that year just happened to be a steam engine, Milwaukee Road's restored Northern type No. 261, built by Baldwin in 1940 and restored for excursion service in 1993.

It seemed only appropriate that the town's annual summer festival should be railroad oriented. The celebration is named after the founder of the Great Northern Railroad, James J. Hill, the builder of empires who opened Minnesota and the northwestern United States to settlement in the late 1800s.

Hill also played a large part in the development of Wayzata in the 1870s and 1880s. "This was the 25th annual James J. Hill Days, and we wanted it to be a special event," said Diane Wilson, director of the Wayzata Chamber of Commerce. The Chamber is headquartered in the former Great Northern depot that was granted to the city by the Burlington Northern in 1976. It didn't take community leaders long to realize that the steam train would be a memorable highlight of the street fair. Milwaukee Road's 4-8-4 locomotive is stored and maintained right in Wayzata's backyard in Minneapolis.

Left: In Wayzata, Minnesota, on September 12, 1999, Milwaukee Road No. 261 is readied for another trip out to Howard Lake. The restored Northern type 4-8-4 helped the city of Wayzata celebrate James J. Hill Days.

Right: A logo is embedded into the front of the old Gallatin Gateway Inn in Salesville, Montana. Restored to its former glory, the Gallatin Gateway Inn served as a hotel in the 1920s and 1930s for Milwaukee Road passengers venturing west to Yellowstone National Park.

LEFT: The restored Northern type locomotive, Milwaukee Road No. 261, thunders past as it leaves Wayzata, Minnesota, to Howard Lake, and back as part of the James J. Hill Days celebration on September 12, 1999.

ABOVE: Crewmen on the Milwaukee Road No. 261 wait for passengers to board the train in Wayzata, Minnesota, on September 12, 1999.

The excursion trains ran twice daily on the Saturday and Sunday of the festival. The morning run left Wayzata, traveled eastbound to Minneapolis Junction, and returned after a roundtrip of 30 miles. The afternoon's run took the westbound train out to Howard Lake and back, a 60-mile roundtrip. Because the train could not be turned, a new BNSF GE C44-9W diesel provided power at the end of the consist for the pull back to Wayzata. The diesel was coupled at the end of the train to the former–Milwaukee Road sky-top lounge, *Cedar Rapids*, owned by the Friends of the 261.

Between excursions on Sunday afternoon, No. 261 shared the spotlight with a parade down Lake Avenue. The local high-school marching band led the parade and, between musical pieces, imitated the sound of a diesel locomotive horn with their instruments. Antique bicycle clubs, a steam tractor called the *Swedish Snowmobile*, and a local dog club were just some of the participants in an eclectic Minnesota parade. Garrison Keillor would have found good material for his radio show out of St. Paul. An impressive sight was the Wayzata Senior Westies Walking Club, which consisted of 100 West Highland White Terriers with red bandanas around their necks. The dogs were on leashes and walked beside their owners.

It was estimated that close to 10,000 people attended the festival each day and at least 2,600 rode the train over the four excursion runs.

RIGHT: Milwaukee Road No. 261 steams past one of the many lakes and ponds dotting the countryside west of Wayzata, Minnesota, on September 12, 1999. The excursion train is en route to Howard Lake on her last afternoon trip of the day.

It would be hard to find a more perfect setting for a street fair celebrating railroading. The old Great Northern mainline, now the BNSF, runs parallel to Lake Avenue, Wayzata's main street. Upscale shops and restaurants face the lake as they line the picturesque street. Sailboats and classic cruisers dock at a pier near the depot.

In the 1890s Wayzata was a playground for the wealthy people of Minneapolis–St. Paul. Great Northern trains brought families seeking sport and relaxation to the many hotels and cabins in Wayzata and along Lake Minnetonka. Hotels, like the Lafayette and Del Otero, were grand hotels with wide verandas to catch the lake breezes.

Passenger steamships plied the clear waters of the lake. James Hill owned many of the ships and lodges on Lake Minnetonka. He also invested heavily in farms and experimental stations, where farmers searched for ways to raise cattle and grow crops that would be resistant to drought.

The hotels are gone today. Most of these grand wooden structures burned in fires. The Del Otero was one of the last surviving until 1948. Wayzata serves now as an upscale bedroom community for Minneapolis–St. Paul.

LEFT: Milwaukee Road No. 261 leaves Wayzata, Minnesota, under a cloak of steam and smoke on September 12, 1999, as part of the James J. Hill Days celebration.

261
261

Placement of the tracks along Lake Avenue originally put the community at odds with James Hill. Charlie Schoene, a local Wayzata rail historian, thinks the citizens of his town just got "too big for their britches" in the 1890s. In 1891 the city filed suit against Hill and the Great Northern. The city wanted the tracks moved away from the avenue because of the noise of railroad switching at night.

Wayzata lawyers claimed that the Great Northern broke a state law concerning the distance railroad tracks must be from a body of water. Hill stated that if he lost the legal dispute, he would move the depot a mile west of town for the next 20 years.

Hill lost the suit and moved the depot. Schoene thinks Hill was unjustly accused of being responsible for the railroad placement. The tracks were laid in the position through Wayzata in 1867, before Hill assumed leadership of the old St. Paul and Pacific, precursor to the GN in 1869. The people of Wayzata still held Hill accountable. But they also began walking 1 mile out of town to catch the trains back to Minneapolis.

Tom Wise, a local boat builder and friend of Hill, convinced the railroad to bring the Great Northern depot back into town in 1906. The 20 years of walking a mile to catch a train and Hill's revenge were brought to an early conclusion.

RIGHT: In Wayzata, Minnesota, engine crews prepare the restored Milwaukee Road No. 261 for her trip to Howard Lake and back on September 12, 1999.

ST. PAUL
AND PACIFIC

UNION PACIFIC
UNION PACIFIC
RAILROAD
CROSSING

8 Union Pacific and North Platte, Nebraska: History Along the Overland Route

Railroads are one of the threads that connect the fabric of the American landscape. Twin shining rails course through valleys and across mountains and prairies, binding together this patchwork quilt of North America. Each town has its institutions, architecture, and history. Train stations across the land once provided the setting for a multitude of personal dramas, as sons went off to war and daughters left for college or careers in the big city.

The Union Pacific across Nebraska is still busy with fast freights at 70 miles per hour. On occasions, the UP sends one of its two operating steam engines out on the line. Most recently UP No. 3985 made a dash across the Corn Husker state pulling the *Super Bowl Special* to Texas in January 2004.

Retracing the route of the UP while photographing its No. 3985, travelers came face to face with one of the nation's first threads in the transcontinental quilt.

The heartland is alive today with stories of the farmers, ranchers, and merchants who comprise the communities on the Platte River. The melodious names of places along Route 30 and the UP mainline called Lodgepole, Ogallala, Kearney, and Cozad recall the history and adventure of the settlement of this region along the old Oregon Trail, the Great Plains, and the Union Pacific lines.

North Platte on the UP mainline is one such historic town, known primarily for the North Platte Canteen and the goodness shown by town residents to American troops passing through on trains bound for World War II.

LEFT: Left: Union Pacific Challenger No. 3985 steams across Nebraska pulling a Super Bowl Special to Houston Texas from Cheyenne, Wyoming, on January 12, 2004.

RIGHT: On a cold morning on January 12, 2004, Union Pacific Challenger No. 3985 passes one of the many grain silos on its dash across Nebraska en route to the Super Bowl festivities in Houston, Texas.

RIGHT: Crews run tests on the 4-6-6-4 Challenger UP No. 3985 in the yards of North Platte, Nebraska, on the morning of January 12, 2004, as the train prepares for departure eastbound to Houston, Texas.

ABOVE: The UP Challenger No. 3985 steams under an overpass as the train Super Bowl Special rumbles across the landscape of Nebraska in the dead of winter on January 11, 2004.

OVERLEAF: Can there be a more stirring sight than the Union Pacific Challenger UP No. 3985 at full stride across the Nebraska landscape on her home rails of the Overland Route, January 12, 2004.

The North Platte Canteen happened almost by accident. The attack on Pearl Harbor had recently occurred and lingered fresh on Americans' minds when North Platte citizens turned out at the depot to greet their boys from Company D of the Nebraska National Guard. The unit had been activated for overseas service and was thought to be on a UP train coming through North Platte. When the train arrived, well-wishers gave baskets of food and snacks to the soldiers. There was only one problem; the train carried the troops of Company D from Kansas. But the 500 or so people who gathered were encouraged by the good feelings of doing something for their country's men and boys headed off to war.

One young woman, Rae Wilson, felt so inspired by the event that she wrote to the local newspaper, encouraging the formation of a canteen at the UP depot. She suggested that food and drinks be given to all servicemen as their trains paused for a 10-minute service stop of the ubiquitous steam locomotives. William Jeffers, Union Pacific's president and a North Platte native, gave speedy approval of the proposal for the canteen to occupy one end of the depot.

The North Platte Canteen opened for business on Christmas Eve 1941. From around 5 a.m. until past midnight everyday, legions of volunteers, mostly women, manned the welcome center in the old brick depot.

3985
U.P.
4-6-6-4-4-69

985

Above: Union Pacific No. 3985 wastes no time in leaving North Platte, Nebraska, just after 8 a.m. on the morning of January 12, 2004.

Left: Early on the morning of January 12, 2004, a young resident of North Platte, Nebraska, who appreciates vintage machines brings his Indian motorcycle down to the UP mainline to get a look at the vintage steam locomotive No. 3985.

Using precious supplies of eggs and sugar, the people of North Platte served America's bravest and best as the troops passed through their town. Each day 4,000–5,000 troops received sandwiches, snacks, and encouragement. The service continued for the next five years.

More than one romance was sparked between a soldier and one of North Platte's young unmarried women serving in the canteen. Letters were exchanged, and a number of postwar marriages resulted, all due to a few minutes in North Platte.

Throughout the various theaters of World War II, soldiers would talk about home and often said something like, "Hey, did you go through North Platte? Man, what a place! I'll always remember that stop." Of all the United Service Organization (USO) centers around the country during that time of war, none would top the North Platte Canteen.

Passenger service through North Platte ended in 1971 with the coming of Amtrak. No longer would trains with names like the *City of San Francisco* and *Challenger* pass through North Platte. Regular passenger service was gone from the Overland Route.

The UP depot was used by freight service for a while. Then one day in 1973, the wrecking ball appeared. In a time of an urban renewal mentality, many historic places became only memories. But around the country today, people of the World War II generation still remember the North Platte Canteen.

9 Changes Along the Rio Grande Narrow Gauge: the Durango & Silverton

In the 1880s, while lumber companies pushed their logging railroads into the last remaining stands of virgin timber in the country, mining companies in Colorado explored every mountain canyon west of Denver in search of gold and silver. Silver proved to be king in the mountains of southern Colorado, and the new railroad under construction by General William Jackson Palmer provided the transportation out of the wilderness for millions of dollars in mineral wealth.

From 1879 to 1881, Palmer built the San Juan extension of his Denver & Rio Grande (D&RG) Railroad over Cumbres Pass to Chama, New Mexico. The narrow gauge line then charged west to the Animas River and finally reached the town of Durango, Colorado, in August 1881.

The silver mining camp of Silverton, Colorado, up the Animas River, needed a railroad. Ore had to be transported by burro and on foot along precipitous trails. An obstacle blocked the railroad's path to Silverton. How do you build a roadbed along a 400-foot sheer cliff through the canyon of a rapids strewn river? Palmer knew that his plans for dynamiting a platform for the railroad bed along such a seemingly impossible route would be met with skepticism by his backers in Denver, so he proceeded without telling his financiers. The tracks, an engineering marvel, finally reached Silverton in July 1882.

RIGHT: No. 478 takes the turntable in the yards of the Denver & Rio Grande Western (D&RGW) narrow gauge in Durango, Colorado in 1979.

FAR RIGHT: The morning D&RGW Silverton steams past the water tank west of Durango, Colorado, in summer 1979.

HERMOSA
CREEK

473
473
473

ABOVE: Locomotive No. 478 waits in front of the Durango depot. The train is ready to board passengers for the trip to Silverton. The narrow gauge railroad has been sold to Charles Bradshaw and is now the Durango & Silverton Narrow Gauge Railroad (D&SNGRR).

LEFT: In June 1979, D&RGW No. 473, a 2-8-2, simmers in the roundhouse in Durango, Colorado, waiting her call for the next passenger train on the narrow gauge line.

Silver mining boomed until 1893, when Congress repealed the Sherman Silver Purchase Act. This law had required that the U.S. Treasury buy more silver than needed for coinage purposes, but with its repeal, silver prices dropped like a rock, losing close to half its value. Shipments from the mines around Durango declined and some operations closed entirely. With a major source of revenue diminished, the narrow gauge railroad had to adapt to the changing times. Tourism would become one of Colorado's major industries in the next 100 years, and the Rio Grande Railroad lead the way in bringing greenhorns to the West.

Excursion trains existed from the beginning of the Rio Grande narrow gauge, and the trains often stopped for passengers to gather wildflowers, fish the streams, or have mountain meadow picnics.

The Rio Grande aggressively pursued marketing of the wonders of its railroad and the majestic mountains through which it passed. Whereas standard gauge railroads in the west, like the Santa Fe and Great Northern, promoted the Indian cultures and lands through which their rails passed, the Rio Grande touted simply the beauties of the Colorado Rockies. The Rio Grande's logo, "Scenic Line of the World," said it all.

Tourists came to the narrow gauge country of Colorado in increasing numbers during the late 1890s. For adventuresome passengers, completing the narrow gauge circle tour became one of the ultimate trips, perhaps akin to the challenges of today's reality television extreme race programs. Circling the southern Rockies around the Rio Grande narrow gauge empire took several days of travel and, in its last iterations, was a true step back in time. Open platform observation coaches, pot-bellied stoves for heat, and coal oil lamps for light showed travelers in the 1940s that they were in the wilds of Colorado on the *San Juan*, not back east on the *20th Century Limited*.

The narrow gauge circle tours ended when a major slide closed a section of the Denver & Rio Grande Western's (D&RGW) Rio Grande Southern section over Marshall Pass. The last version of the *San Juan* passenger train on the D&RGW ran in 1951. Passenger service disappeared from the narrow gauge system, except for the Silverton run from Durango that continued to draw tourists from all over the world. The D&RGW' Silverton branch also appealed to Hollywood movie companies.

Films like *Around the World in 80 Days*, *A Ticket to Tomahawk*, *How the West Was Won*, and *Denver and Rio Grande* found locations in and around the railroad line at Durango for their productions. Probably one of the most popular movies ever filmed on the D&RGW was *Butch Cassidy and the Sundance Kid* in 1969.

RIGHT: The morning train to Silverton on the Durango and Silverton Narrow Gauge Railroad rounds a curve west of Durango, Colorado, in 1981.

DURANGO & SILVERTON
478

Through the 1950s, the Silverton excursion train grew in fame and ridership, but freight on the D&RGW from Alamosa to Durango continued to decline. A brief oil field boom in Farmington, New Mexico, brought a reprieve for the southern segment of the old narrow gauge circle. Pipe and supplies moved by rail over Cumbres Pass to Chama, New Mexico, and westward to Durango.

Eventually even this oil field business dwindled, and in 1967 the D&RGW filed for abandonment of the narrow gauge railroad. In 1969 the states of Colorado and New Mexico began putting together the two-state partnership that resulted in the Cumbres & Toltec Scenic (C&TS) Railroad in Chama. The popular Silverton passenger train continued in operation under the D&RGW. The rails between Chama and Durango were removed.

The D&RGW sold its narrow gauge empire in Durango in 1981 to Florida businessman, Charles Bradshaw. The new railroad changed name to the Durango & Silverton Narrow Gauge (D&SNG) Railroad. Bradshaw seemed good for the railroad. The D&RGW bid farewell as a corporate entity in the Animas River valley after 100 years. Ownership of the railroad changed again in 1998 when Allen and Carol Harper purchased the D&SNG.

Below Right: Locomotive No. 478, a 2-8-2, waits in Durango, Colorado, for a morning departure on the newly purchase narrow gauge line now called the Durango & Silverton Narrow Gauge Railroad.

Below: In the roundhouse in Durango, Colorado, No. 473 waits for the call to pull a passenger train to Silverton. The D&RGW narrow gauge line is nearing its 100th anniversary and will soon be sold to Charles Bradshaw in 1981.

Durango grew and has become a tourist destination over the last 50 years. Whitewater enthusiasts and mountain bikers flock to Durango throughout the year. The nearby ski area offers world-class alpine skiing. Narrow gauge steam engines pull passenger trains out of Durango on a daily basis, year round. The new D&SNG offers backpackers, who seek to explore the wilderness of the San Juans, a great ride to the front door of mountain meadows and high peaks.

ABOVE: A worker lubricates one of the vintage narrow gauge steam locomotives of the D&RGW Railroad in the Durango, Colorado in June 1979.

10 Rough Times on the Reader: Tales of an Arkansas Shortline

Like mushrooms sprouting on a lawn overnight, logging railroads appeared all over North America during the 1880s. Typically these railroads laid light rail with minimal ballast and ventured into the deep woods to bring in the remaining virgin timber. The owners of Reader Mill, operating in Arkansas' counties of Nevada and Ouachita, saw the need for a locomotive to replace its oxen teams in hauling logs. The sawmill purchased a small steam engine, and the railroad that became the Reader was born.

After going through several different owners, the logging railroad became the property of Mansfield Hardwood Company in 1923. The town of Reader, Arkansas, rose from the wood chips and sawdust near the Sayre post office on the new St. Louis, Iron Mountain & Southern Railroad branch from Gurdon to Camden, Arkansas. The new town was named after the first owner of the sawmill, Lee Reader. Changes loomed on the horizon for the small lumber hauling railroad.

Drillers discovered oil in the region of Waterloo, Arkansas, in 1921. There was only one problem: the oil was so thick that it was considered asphalt and couldn't be pumped through a pipeline. Transport of the asphalt required rail tanker cars.

Right: Reader No. 2, a Mogul type, is under steam at Reader, Arkansas. This engine is currently undergoing renovation by Richard Grigsby, owner of the railroad.

Below: Reader No. 7, a Prairie type locomotive, waits on platform at Reader, Arkansas for the next train to Camp Dewoody in spring 1979.

2

BALDWIN LOCOMOTIVE WORKS
7
PHILADELPHIA, U.S.A.

LEFT: Reader No. 7 crosses one of the many trestles over Caney Creek out of Reader, Arkansas. The train is bound for the end of the line at Camp Dewoody.

The Mansfield Lumber Company saw an opportunity and chartered its logging operation as a common carrier railroad in 1925.

With the formation of the Reader Railroad as a common carrier in 1925, the interchange of oil tanker cars occurred at Gurdon on the Missouri Pacific Railroad, a descendent of the old Iron Mountain line. Berry Refinery's asphalt traffic kept the Reader in business, even as logging cycled through financial peaks and valleys.

Through the 1940s and 1950s the Reader Railroad continued ambling through the tall trees, hauling logs to the sawmill and asphalt from the oil wells. In 1956, Mansfield Hardwood Company sold the Reader Railroad to Tom Long.

Long was an attorney from the University of North Carolina who served as manager of the Mansfield Lumber Company's retail lumberyards. Some friends teased Tom Long that he had actually bought a model railroad scaled 12 inches to the foot.

BELOW: The switch is thrown at Camp Dewoody as Reader No. 7 has been turned and will pick up its train for the return to Reader, Arkansas, in the spring of 1979.

BELOW RIGHT: The fireman of Reader No. 7 checks his fire in locomotive No. 7 on the Reader Railroad in southern Arkansas in the spring of 1979.

David P. Morgan, editor of *Trains* magazine, found Tom Long's Reader Railroad in the piney woods of Arkansas irresistible. At a time when Morgan and his friend, photographer Phil Hastings, wrote about their search for the last vestiges of steam railroading in Canada, the Reader continued its low-key operation without diesel engines. The Reader eventually became "the last regularly scheduled mixed train drawn exclusively by steam locomotives," according to the *Official Guide of the Railways* of May 1966.

In April, 1966, Morgan and his wife came down to Reader to help Tom Long christen the line's newest acquisition, Consolidation No. 1702. The locomotive had originally been built by Baldwin for the U. S. Army Corps of Engineers for use in Europe in World War II. It never saw overseas service and, in 1946, went to another Arkansas shortline, the Warren & Saline River Railroad. Reader became the locomotive's second owner in 1964, and the engine underwent renovation to become the star attraction of the Reader.

As railfans throughout the country read David Morgan's Trains magazine pieces about the Reader, more people showed up each year to ride the mixed trains, which passed the long vanished settlements along the shortline with names like Goose Ankle, Cummings Springs, and Possum Trot. The Reader's nickname is the "Possum Trot" line because the land is so swampy, thus it is "fit only for possums to trot on."

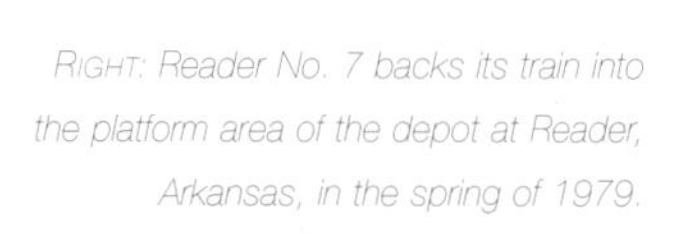
RIGHT: Reader No. 7 backs its train into the platform area of the depot at Reader, Arkansas, in the spring of 1979.

READER

7
READER
READER

RIGHT: In the spring of 1979, the fireman of Reader No. 7 picks up another piece of wood for the Prairie type locomotive brought to the Reader Railroad from the Victoria, Fisher & Western Railroad in Louisiana.

LEFT: Reader No. 7, a 2-6-2, Prairie type built by Baldwin Locomotive Works, boards passengers at the end of the line on the Reader Railroad at Camp Dewoody in the spring of 1979.

By 1973 tourist passengers and logging had declined on the Reader Railroad. Berry Refinery closed its plant. The transport of asphalt had given the Reader its real reason for existence since the shortline's inception. Tom Long began removing track and the line appeared doomed to abandonment. Richard Grigsby, whose family had long been in the timber industry in the area of Malvern, Arkansas, lead a group of investors to purchase the line in 1975.

Grigsby and his group brought several locomotives up to Arkansas from the logging woods of east Texas and Louisiana in 1973. At Reader, these included the No. 2 and No. 7, which still remain on the railroad today. No. 7 is a wood burner with a large diamond stack. Throughout the 1980s, No. 7 provided the power for trains that carried passengers from Adams Crossing, out to Camp Dewoody, and back.

In its latter day operation, the Reader continued until the early 1990s. The shortline is currently out of service. A few years ago even the community of Reader, Arkansas, abandoned its status as an incorporated town.

11 Life on the Narrow Gauge in Chama, New Mexico

Going to Chama, New Mexico, is like traveling via a time machine. Imagine the world before cell phones, iPods, the internet, fast food, CT scanners, jet airplanes, automobiles, penicillin, and even electric lights. Stroll through the railyard and shops of the Cumbres & Toltec Scenic Railroad in Chama and you enter a world that began before any of these technologies and products existed.

The railyards of Chama have been in place for a long time. The Denver & Rio Grande built through Toltec Gorge and over Cumbres Pass in 1880 as part of the San Juan Extension to tap mining potential in the southern Rocky Mountains. The railroad reached Chama in 1881 and proceeded west to Durango, Colorado, and eventually completed a loop around the heart of the southern Rocky Mountains.

The D&RG and its successor, the Denver & Rio Grande Western (D&RGW) operated the narrow gauge railroad through Chama until the 1970s, when the historic line was bought by the states of Colorado and New Mexico. The resulting historic park, leased to individual operating companies through the years, survived as the C&TSRR.

The times have changed, but through the summer months, Chama carries on with 100-year-old narrow gauge rituals on a daily basis. These acts of railroading remind us of life in the southern Rockies long before cell phones and McDonalds came into our world.

ABOVE: Cumbres & Toltec Scenic Railroad No. 487 has backed down to the water tank to top off the tender for the day's run up Cumbres Pass.

RIGHT: C&TSRR No. 489 storms past an ancient Rio Grande water tank on the climb to Cumbres Pass.

489
489
489

John Bush was passionate about steam railroading. He left Chama to go to the White Pass and Yukon for a time and currently manages the Roaring Camp Railroad in northern California. During the 1990s he served as chief mechanical officer for the C&TS.

In a 1993 interview in Chama, Bush discussed the New Mexico narrow gauge. "What is interesting about this place," he said, "is that it genuinely functions. It runs a lot of trains, has a lot of locomotives, and is all still complete. The locomotives take water where they always took water, they get coal and sand at the same facilities they always have. There is a depth and continuity here that you don't get at other places."

In the shops that Bush managed in the mid-1990s, there emerged a picture of three types of workers on the old narrow gauge line. Firstly, there were the local people, some in their 20s, who had fathers and grandfathers who worked for the D&RGW back in the 1940s and 1950s.

BELOW: John Coker, engineer for the C&TSRR, oils around Mikado No. 488 in Chama, New Mexico, before the morning departure for Antonito, Colorado, in 1993.

RIGHT: Early in the morning in Chama, New Mexico, narrow gauge locomotive No. 489 on the C&TSRR moves to the ash pit to drop the night's ashes before hooking up to the morning train bound for Antonito, Colorado.

Secondly, there were workers who had trades useful to a steam locomotive shop and came up from Santa Fe or Albuquerque just because of the availability of a good job. Finally, there were young people who became enamored with the lore of narrow gauge railroading in the Rocky Mountains and wanted nothing more than to be around live steam engines for a season in their life. All three types of workers shared one common trait. They seemed to care about the C&TS and the part they played in working on a railroad that is a survivor from another time.

ABOVE: In the early morning hours in Chama, New Mexico, No. 484 backs into position to take water from the double spout water tank in preparation for the run up Cumbres Pass to Osier, Colorado.

489
D&RGW
K-36

The yards and shops of the Cumbres and Toltec have always seemed friendly and open. As long as people observed commonsense safety rules, visitors could go just about anywhere they wanted.

Stepping down the embankment from the main road through Chama, visitors were met by the smells of coal smoke, steam, hot grease, and creosote. A stall from the old brick roundhouse survived on the C&TS, but most repair work was done in the newer shop building attached to the vintage building.

A hammer rang loud on cold metal and echoed throughout the locomotive shop as a worker pounded on a locomotive part. The shop smell always seemed familiar and nostalgic, whether it was a blacksmith's operation on your grandfather's farm or a steam locomotive shop in the mountains of New Mexico. In Chama there was that mixture of the smells of hot grease and burning coal, which took one back to childhood days in a rural setting.

Track crews prepared for the day's run up the line toward Antonito. They left after the morning train departed, around 10 or 10:30 a.m. D&RGW trains, like the *San Juan*, came through Chama bound for Durango at about the same time in the 1920s. Time moves on, but some things have remained the same.

LEFT: C&TS RR No. 489 picks up speed climbing out of the Chama River valley toward Cumbres Pass in summer 1993.

484

ABOVE: The narrow gauge excursion train leaves Antonito, Colorado, on the Cumbres and Toltec Scenic Railroad in 1993.

LEFT: No. 484 on the Cumbres & Toltec Scenic Railroad takes a long drink from the water tank in the rail yards of Chama, New Mexico, on a summer morning in 1993.

John Coker was a railroader from 1989 until 1995. He was born in Los Angeles, California, and moved east to Denver where he completed two years of college. Whereas many undergraduate students in Colorado discovered the joys skiing Breckenridge and Aspen, Coker found the fascination narrow gauge railroading. He began working for C&TSRR in 1973.

Coker ran trains on the narrow gauge and spent the long winters painting scenes from the summers. His skills improved and his work appeared on the covers of magazine. He began selling his original work and prints. Coker railroaded on both the Durango & Silverton Narrow Gauge Railroad and the Cumbres & Toltec Scenic Railroad. Finally in 1995, he retired and now lives in Bayfield, Colorado, where he devotes his time to producing and marketing his art.

Chama, New Mexico, continues luring artists, mechanics, and romantics like John Bush and John Coker to the high country of the narrow gauge railroad. They have become enchanted by the time travel that the old Denver & Rio Grande Western still provides.

12 The Cumbres & Toltec Scenic Railroad: Battles with Nature on Cumbres Pass

William Jackson Palmer might have chosen another route for his narrow gauge empire reaching south from Antonito, Colorado, toward Durango and Silverton. In 1879, he had several other choices, one of which was lower in elevation but longer in miles. Palmer chose Cumbres Pass for the path of his newly fledged Denver & Rio Grande (D&RG) Railroad. Palmer and the D&RG soon became acquainted with the tribulations that Rocky Mountain winters could foist upon railroad operations in that part of the country.

Construction of the narrow gauge railroad slowed during the winter of 1879, and track building through Toltec Gorge was stopped completely for a time due to heavy snows. The winter of 1880 was even worse, but construction continued over Cumbres Pass and the Continental Divide. Finally in January 1881, the D&RG reached Chama, New Mexico, a future center for operations in the coming years.

Operations out of Chama across Cumbres Pass became more difficult with each year. At first, work crews were called from Chama and the line was cleared by hand with shovels. In the late 1880s, Orange Jull of Ontario, Canada, designed a rotary snowplow. His patents and ideas were subsequently obtained by a manufacturing firm lead by two brothers, John and Edward Leslie. The Leslies perfected the rotary snowplow and contracted the building of rotary snowplows by the Cooke Locomotive and Machine Works. Eventual consolidations of these companies found all snowplows built by American Locomotive Company after 1903.

Left: The Cumbres & Toltec Scenic Railroad Rotary OY clears the narrow gauge line on May 4, 1991.

Above: The work train and snowplow have cleared a path up the narrow gauge railroad on May 4, 1991, in a special train to open the narrow gauge tourist line for the new season.

Clearing Cumbres Pass by hand with shovels and pure manpower proved arduous. Passenger trains became stranded on the pass by blizzards for days, and evacuation often required sleds and skis. The heavy snowstorms of the winter of 1883–1884 set a benchmark for the challenges of railroading over Cumbres.

Finally in 1889, the Rio Grande's first rotary snowplow came from Cooke Works and eased the burden of clearing the narrow gauge line. That machine carried the designation of Rotary No. 1, but was changed to Rotary OM when other plows began arriving. Rotary OM survives today on the C&TS' operation of the Chama to Antonito line.

Three more snowplows came to the Rio Grande in the years after 1890. Rotaries OO and ON were both built by Cooke and lasted until the 1950s, but were eventually scrapped. American Locomotive Company acquired the Cooke Works and delivered the last snowplow, Rotary OY, to the Rio Grande in 1923. Rotary OY has continued in intermittent operations in Chama to the present time.

BELOW: At Cumbres Pass, a crewmember on the narrow gauge work train pulls the waterspout away from the tender of the Mikado steam engine, providing pushing power for the Rotary OY. The snowplow has worked for most of the day on May 4, 1991, to clear the line to Cumbres Pass. Snow is beginning to fall as evening arrives in the Rocky Mountains.

ABOVE: Mikado No. 488 provides the push for Rotary OY as the line is cleared to Cumbres Pass on the C&TSRR on May 4, 1991.

By the 1920s, the Rio Grande purchased its K-36 and K-37 locomotives that weighed close to 90 tons each. When equipped with a pilot plow, these Mikados could clear snow on the tracks up to 6 feet in height. When the snow pack exceeded this depth, the rotaries were called out.

Both Rotaries OY and OM were used on the D&RGW railroad through the 1960s. As traffic declined on this narrow gauge railroad in the late 1960s, operations over Cumbres Pass ceased during the winter. Freight was diverted to the highways and the railroad was cleared with backhoes and other heavy equipment. The rotary snowplows sat quiet in the yards in Chama and Antonito.

ABOVE: A blizzard fills the air with flying snow as Rotary OY and Mikado No. 488 clear the narrow gauge mainline on the C&TSRR below Windy Point on May 4, 1991.

RIGHT: C&TSRR Rotary OY has just blasted through a drift on the narrow gauge line on May 4, 1991.

ABOVE: On May 4, 1991, a work train with Rotary OY clears the narrow gauge line approaching Windy Point just below Cumbres Pass, Colorado.

In the winter of 1957 President Dwight Eisenhower lead the country and Elvis Presley wowed the girls. It was also the time when Rotary OM received its last call to rescue a train stranded on Cumbres Pass for the D&RGW. The rotary plow and three locomotives from Chama pushed toward Windy Point and cleared one trapped train.

Another train remained encased in snow and the snowplow crew proceeded. However, Rotary OM became trapped by a snow slide behind it and this predicament required Rotary OY's intervention on the hill. This drama lasted for over a week and saw the eventual rescue of the train crews by helicopters and U.S. Army snow machines.

By 1969, the D&RGW decided to abandon its narrow gauge line from Antonito to Durango. Concerned citizens of Colorado and New Mexico came forward and forged a two-state compact with state funding to support the 56 miles of railroad as a tourist operation. The line from Chama received preservation, but the rails from Chama west to Durango were eventually removed.

In 1971, the Cumbres & Toltec Scenic Railroad became operational. Scenic railways evolved as the operational entity for the new tourist line in the mountains. In an effort to spur interest in the narrow gauge shortline, the C&TS began operating snowplow Winter Specials. On January 12, 1974, Rotary OM was brought back to service and fired for the first time since the blizzard in 1957. The Winter Specials proved so popular that these runs were repeated annually for several years.

SECTION MEN
CUMBRES & TOLTEC SCENIC R.R.
488

ROTARY
OY
C & T S

LEFT: Rotary OY on the C&TSRR throws a plume of snow from the narrow gauge tracks on May 4, 1991.

Through the years since 1971, the C&TS narrow gauge line and it rotary snowplows have been used several times for movie and television productions. In March 1979, a film company settled into Chama for two days of shooting for a Miller Beer commercial to air in the next year's Super Bowl football game. The event had not been publicized at the production company's request. The film crews didn't want railfans with their cameras cluttering the background of their commercial. If you looked closely when the commercial aired, you could still see a certain set of ski tracks down the center of the tracks as the train passed by Juke's Tree at the Chama River Bridge.

In 1982, Kyle Railways assumed operation of the C&TS. In 1991, the railroad wished to expand its season and begin operations by Memorial Day weekend. Heavy snow throughout the spring blocked the line. In cooperation with the Friends of the Cumbres & Toltec, a support group for the shortline, Kyle Railways planned a clearing of line using Rotary OY.

Railroad photographers from around the country came to the snowy mountains of Chama on May 4–5, 1991. Joining them were video companies and a CBS television news team with Bob McNamara. Through blizzards alternating with sunshine on May 4, and clear skies on May 5, Rotary OY and her work train offered images reminiscent of the 1920s.

BELOW: The caboose of the work train behind Rotary OY pauses at Cumbres Pass, Colorado, before returning to Chama, New Mexico, for the night on May 4, 1991.

D
WESTERN
1218

13 Norfolk & Western No. 1218 and a Conductor Remembered

When railroad photographers grow old and their great grandchildren gather at their feet, those old timers who shot slides and black-and-white film will probably tell of the golden age of steam railroading, the 1980s. Looking back, the 1980s does seem like the good old days.

Out west the Union Pacific looked for any opportunity to put its preserved Northern No. 8444 or restored Challenger No. 3985 out on the road. Southern Pacific No. 4449 had just steamed down to Sacramento, California, for the opening of the California State Railroad Museum in 1981. Steam renovations were sprouting everywhere. The excursion schedules in *Trains* and *Railfan and Railroad* magazines determined vacation plans for the coming summer for many families. In the 1980s, a Pennsylvania Railroad K-4 had returned to service for a season, and Ross Rowland Jr. delighted steam fans with his activities in the northeast.

There were two active steam engines running across the Norfolk Southern system each summer by the late 1980s. These locomotives provided opportunities for classic photography for southern railfans. Norfolk & Western (N&W) No. 611 steamed back into life in 1981 and pulled a number of annual excursions across the newly expanded Norfolk Southern system. In 1987, N&W No. 1218, an articulated 2-6-6-4, was restored by the railroad's steam shops in Birmingham, Alabama, and joined No. 611 in active service.

LEFT: Norfolk & Western (N&W) No. 1218 is on the road to Toccoa, Georgia, in the spring of 1988.

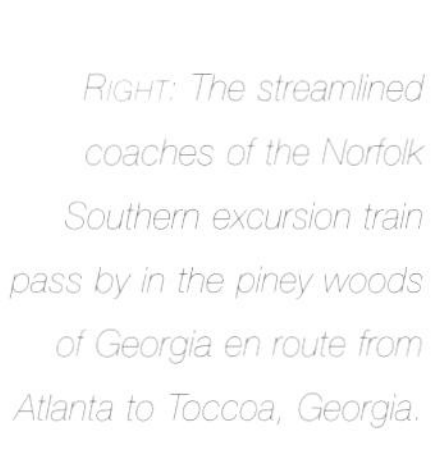

RIGHT: The streamlined coaches of the Norfolk Southern excursion train pass by in the piney woods of Georgia en route from Atlanta to Toccoa, Georgia.

Right: N&W No. 1218 pauses for running gear check on the excursion train between Atlanta and Toccoa, Georgia, in the spring of 1988.

Above: An overpass on the road between Atlanta and Toccoa gives a look at details of N&W 2-6-6-4, No. 1218.

One of the tales old timers may tell their great grandchildren is the story of Norfolk Southern conductor, Harold Duane "Cowboy" Mintz. He became a legend in railroading and worked on N&W No. 1218 excursions several times.

Newcomers to Norfolk Southern excursions out of Atlanta might have been surprised when the conductor showed up wearing the traditional coat and tie and a black western cowboy hat. Mintz and his black hat with the metal conductor's badge on the crown made a striking picture. His gray beard and erect bearing resembled a civil war general more than a railroad conductor.

Cowboy Mintz grew up not far from the tracks of the Southern Railroad in Rockmart, Georgia. According to his friend, Lamar Wadsworth, Mintz' father operated the city's water plant and his grandfather ran a small industrial locomotive at the local cement plant.

When he was a kid, Mintz loved horses and rode them frequently in the family pastures beside the tracks. He would race the trains and wave at the conductors and engine crews. Wadsworth related that Mintz also rode in rodeos for a time. But rodeo life was not where Mintz found his true calling. He loved those iron horses.

At age 18, Mintz became a railroader. He came to work that first day at Inman Yard in Atlanta, and one of the conductors recognized him and said, "If it ain't the little cowboy." This nickname stuck with him throughout the years.

1218

Left: N&W No. 1218 runs on a Norfolk Southern Railway excursion train between Atlanta and Toccoa in the spring of 1988.

Above: The articulation of N&W No. 1218 is apparent in this picture in Chattanooga, Tennessee, in the spring of 1988.

Cowboy Mintz became almost totally deaf by the time of his retirement. He always blamed the start of his hearing loss on one night on the Southern Railway. He was asked to go out on the running board of the steam engine and rap on the sand dome and pipes to get the flow of sand started again. As the young Mintz carried out his duty, the engineer blew the whistle for a crossing. His hearing deteriorated after that.

Lamar Wadsworth, a Southern Railway history buff and resident of Rockmart, befriended Mintz in 1990 after Cowboy's retirement. Wadsworth related that Cowboy Mintz knew as much about Southern Railway history as any one he had ever known.

When Cowboy Mintz retired at age 61 years, he was senior conductor on the Georgia division of the railroad. Mintz spent the last years of his life in Lindale, Georgia, with his family and died at age 73. He was buried at Rose Hill cemetery in Rockmart, not far from the old Southern tracks where he once raced horses against steam trains.

The Bible says there is a season under heaven for everything. There is a time to be born, a time to plant, a time to harvest, and a time to die. After just seven years in excursion service, time ran out for N&W No. 1218. A switching move damaged some of the railroad's excursion coaches in 1994, and Norfolk Southern Railway called an abrupt end to its vintage steam program. The expenses to repair the coaches and rising liability insurance costs for railfan trips sounded a death knell for continued existence of the program. A sad day came to the community of railroad preservationists around the country when the steam shops were closed in Birmingham, and all of the equipment used for maintenance of steam engines was sold.

Norfolk & Western No. 1218 was born in the N&W shops in Roanoke, Virginia, in 1943 and with her retirement the locomotive was placed in storage by the Norfolk Southern Railroad. In 2001, the railroad announced plans to donate N&W No. 1218 to the city of Roanoke in honor of N&W photographer, O. Winston

Below: N&W No. 1218 is an articulated 2-6-6-4.

Link. Eventually No. 611 and No. 1218 were backed under protective buildings for permanent display at the Virginia Museum of Transportation in the city not far from the shops where both locomotives were built.

In 2002, the former N&W depot in Roanoke became the new O. Winston Link Museum, where prints and papers of this famous photographer are displayed. Link documented so well the life of the N&W through his landmark photography in the last days of steam in the 1950s. It seemed right that the two steam locomotives and Link's photography celebrating the life of the N&W came to rest in Roanoke.

Above: N&W No. 1218 runs at track speed through the piney woods of Georgia, from Atlanta to Toccoa, on an excursion in the spring of 1988.

4449
CS-4
SF
MEN
AT
WORK
2

14 Southern Pacific No. 4449: Daylight to the Fair

Left: Men, and probably some women, also are at work cleaning Southern Pacific No. 4449 and her World's Fair Daylight as the train spends the night of June 10, 1984, in San Antonio, Texas, after leaving New Orleans and the 1984 Louisiana World Exposition.

In 1937, Southern Pacific's new Daylight passenger trains inspired hope for the future as the United States emerged from the Great Depression. The SP splashed its newest named trains with a distinctive red, orange, and black paint scheme. The lightweight cars and their streamlined steam locomotives came to be called by many "the most beautiful train in the world." The colors and refinements of Daylight trains conveyed to the country the optimism and bounty of the Pacific coast during a time of suffering in the Great Plains and Midwest.

SP chose Northern type locomotives as the power for their new Daylights. This class of engine carried the designation GS for golden state. Later during World War II, this GS nomenclature was changed to mean general service, with an eye toward persuading the War Production Board to allow the building of more 4-8-4s.

Baldwin built the first 21 GS-1s in 1930. The Cotton Belt constructed the last locomotives of this class, GS-8s, in 1942. Lima Locomotive Works of Lima, Ohio, produced all other Northern types for the SP from 1937–1942. The Lima-built engines carried the designations of GS-2 through GS-7. Most of the 4-8-4s came out of production in Daylight paint schemes, with such streamlined features as wheel skirts, skyline casings, and a distinctive conical smoke-box door. During World War II, some of the Northern types were repainted black and had their streamlined apparel removed.

Below: Southern Pacific No. 4449 has arrived into Houston, Texas, in the evening of June 9, 1984, en route to Portland, Oregon, from New Orleans and the 1984 Louisiana World Exposition.

Below Right: No. 4449 displays its distinctive conical smoke-box door while at rest in Houston, Texas.

4449
4449
5

ABOVE: In 1984, the Southern Pacific depot in San Antonio, Texas, welcomes not only Amtrak passenger trains, but also the World's Fair Daylight, a special train running from New Orleans to Portland, Oregon, powered by SP steam engine No. 4449.

LEFT: Southern Pacific No. 4449, pulling the World's Fair Daylight from Houston to San Antonio, makes a brief stop on the mainline to check the locomotive running gear on June 10, 1984.

During the late 1930s and early 1940s, SP bought a total of 85 General Service Northerns for use in its vast system. With dieselization, most of this class of SP locomotives met the scrapper's torch. By the turn of the twenty-first century, only two SP 4-8-4s remained: No. 4460, a GS-6 in St. Louis, Missouri, at the National Museum of Transportation; and No. 4449, a GS-4 in Portland, Oregon. Southern Pacific donated No. 4449 to Portland, Oregon, in 1956. The locomotive came to rest in Oaks Park alongside another 4-8-4, SP&P No. 700.

The American Bicentennial brought the slumbering SP engine to life when Ross Rowland—a stockbroker and steam enthusiast—chose No. 4449 to lead the *American Freedom Train* as part of America's birthday celebration in 1976. The locomotive was moved from Oaks Park to the Hoyt Street roundhouse in 1974.

Rowland chose a young engineer, Doyle McCormack, to lead the renovation and operation of No. 4449. McCormack came from a family of railroaders in Conneaut, Ohio, not far from where Lima built the SP Northern in 1941. Doyle took a leave of absence from his job with the Norfolk & Western and moved west to Portland.

LEFT: Appekunny Mountain *is an appropriate observation coach for the last car of the* World's Fair Daylight. *The former Great Northern coach has been painted into Daylight colors for the special trip. The steam train is spending the night in San Antonio after running from Houston on the Southern Pacific mainline in Texas.*

In Portland, Doyle McCormack took charge of the restoration of No. 4449. After delays in renovation of the 4-8-4, McCormack and his fellow crewmembers left Portland on June 20, 1975, for their deadhead run to Chicago to meet the *American Freedom Train*. The No. 4449 raced across the country in her new paint scheme of red, white, and blue.

The *American Freedom Train* brought history and culture to the American people in all parts of the country. Included in the display were such diverse artifacts as a copy of the U.S. Constitution, the Louisiana Purchase, Judy Garland's dress from the *Wizard of Oz*, and a moon rock. SP No. 4449 returned to Portland when the *American Freedom Train* run ended on December 31, 1976. The train proved a huge success and became the one Bicentennial event to involve the whole country.

Doyle McCormack returned to his job on the N&W in Ohio, but had Oregon on his mind. "Our horizons had grown immensely," he said. "I started looking around and found that the SP was hiring." In the spring of 1978, McCormack and his wife moved to back to Portland where he became a Southern Pacific engineer.

The now famous Daylight locomotive was basically in storage from mid-1978 until 1981, when the California State Railroad Museum invited SP No. 4449 to its grand opening. For the trip to California, the locomotive and two cars received Daylight colors. With the engine in its original paint scheme again, McCormack and the volunteer group, Friends of 4449, had an idea.

ABOVE: The interior of observation lounge coach, Appekunny Mountain, *appears inviting for the trip to Portland, Oregon, as the train is readied for departure from San Antonio on June 11, 1984.*

The group approached the SP about taking No. 4449 to New Orleans, Louisiana, for the 1984 World's Fair. The railroad gave an affirmative reply and the Friends of 4449 made plans for a Gulf Coast journey. "In 1983, No. 4449 underwent a major rebuild," said McCormack. For this trip, the whole train would be repainted into all Daylight colors. A recreation of the "most beautiful train in the world" emerged and left Portland for a 51-day trip to New Orleans.

McCormack recalls the trip as brutal. "A death march," he said, but he was quick to add, "We had a good time." The train traveled through California, Arizona, and Texas before reaching Louisiana. The trip involved a revolving cycle of three days of travel with one day of rest for the crew. "We were supposed to rest on that fourth day," said McCormack, "but we ended up cleaning the train and doing maintenance on the locomotive."

SP No. 4449 returned to Portland after the *World's Fair Daylight* run and moved into new headquarters at Brooklyn Yard roundhouse.

BELOW: Southern Pacific's No. 4449 hits the road for the journey from Houston to San Antonio, pulling the World's Fair Daylight *on June 10, 1984.*

McCormack and No. 4449 became movie stars in the 1986 movie, *Tough Guys*. Doyle and his crew took the SP engine and a small train down to California for the filming. "We had a great time. They treated us like kings," said McCormack, when asked about the train crew's relationship to the movie production company. McCormack had a small speaking part as the train engineer in the movie starring screen legends, Burt Lancaster and Kirk Douglas.

Numerous other mainline steam locomotives have returned to excursion service in the last three decades. Everyone owes a debt of gratitude to No. 4449 for the renaissance in steam operations that followed the *American Freedom Train* in 1976.

ABOVE: The logo and the colorful paint scheme of the new Daylight train on the West Coast brought hope to a generation of Americans emerging from the Great Depression in 1937. No. 4449 is in Houston, Texas, on June 9, 1984.

15 A *Homecoming Excursion* Up the Columbia River: SP&S No. 700 Returns

The year 1938 found the United States and much of the world still in the stranglehold of the Great Depression. The Dust Bowl of the Great Plains spawned waves of migrants to California and the Pacific Northwest looking for work in the fruit orchards and produce fields. In the midst of those troubled times, the Spokane, Portland & Seattle Railway struggled with secondhand equipment from its parent lines, the Great Northern and Northern Pacific.

The SP&S persuaded company owners to allow the purchase of nine new locomotives: six Challenger types for freight and three Northerns for passenger service in 1937. The Northerns, numbered 700 through 702, were delivered in 1938. The 700s quickly went to work pulling SP&S passenger trains between Spokane and Portland.

Only No. 700 survived dieselization in the late 1940s. When it became apparent that steam was finished, the locomotive pulled a farewell run from Portland to Wishram, Washington, on May 20, 1956. That last run up the Columbia River required 21 cars and carried some 1,400 people.

After the farewell trip, No. 700 was placed in a Portland's Oaks Park, along with SP No. 4449. The two locomotives slumbered in the park for 20 years until each was moved, one at a time, into Brooklyn roundhouse where dedicated volunteers began restorations. No. 4449 was restored first in 1975 and began a tour across the country in 1976 pulling the U.S. Bicentennial celebration's *American Freedom Train*.

RIGHT: Spokane, Portland & Seattle No. 700 pulls the Homecoming Excursion *into Wishram, Washington, on April 20, 2001.*

ABOVE: SP&S No. 700 departs the Amtrak depot in Vancouver, Washington, on April 20, 2001. The train, called the Homecoming Excursion, is the first train to run from Portland, Oregon, all the way to Spokane, Washington, in many years.

RIGHT: A 1938 automobile waits on the side of Washington State Route 14 while the restored SP&S No. 700—built in the same year—accelerates in a steady rain on the run from Vancouver to Spokane on April 20, 2001.

SPOKANE
PORTLAND
AND
SEATTLE RY.
700

Right: On April 20, 2001, SP&S No. 700 steams along the Columbia River on the BNSF mainline along the north shore of this mighty river of the West.

Left: SS&P No. 700 pauses in Wishram, Washington, on the way to Spokane on April 20, 2001. The Homecoming Excursion has left the rain behind in the west and emerged into sunlight along the shores of the Columbia River.

While SP No. 4449 hit the road pulling the *American Freedom Train* across the United States in 1976, Chris McLarney stayed in Portland. He was only 15 years old, too young to join the crew on No. 4449. McLarney and others soon organized the Pacific Railroad Preservation Association (PRPA). Protecting the railroad heritage of Portland and the Pacific Northwest became the group's mission.

In 1989, SP&S No. 700 finally escaped the confines of Oaks Park and moved to Brooklyn roundhouse to join SP No. 4449. Jim Vanderbeck, current president of the PRPA, remembered pounding spikes to lay the temporary track from the park to the tracks of the old Portland Traction Company. Over these rails of this former electric line, No. 700 moved to Brooklyn Yard.

The PRPA set to work restoring No. 700 with the help of volunteer machinists, pipe fitters, and boilermakers. In 1990 the locomotive made its first run to Wishram, Washington, on its historic home line, the Spokane, Portland & Seattle Railway line, now BNSF. During the 1990s, No. 700 made infrequent mainline runs. None went as far as Spokane until April 2001.

The train that made the return to Spokane from Portland in April 2001 was billed as the *Homecoming Excursion*. The "Lady", as No. 700 is called, pulled the passenger train through the scenic Columbia River Gorge passing along the way such geographic icons as Beacon Rock, Cape Horn, and Booneville Dam. Leaving Amtrak station in Vancouver, Washington, on April 20, 2001, in a pouring rain, a caravan of railroad photographers and videographers from around the world pursued the train along Route 14.

Crossing rivers flowing down into the Columbia River from the Cascade Mountains to the north, the train must have seemed like a ghost from the past to fishermen and boaters beneath railway trestles along the way. Somewhere near Hood River, Oregon, the train emerged from the rain and the skies began to clear.

After a morning in the rain and an afternoon in the sun, No. 700 and her train pulled into the tri-cities area of southeastern Washington. Richland, Pasco, and Kennewick, Washington, are agricultural communities with

BELOW: The Homecoming Excursion *has run out of the rain found back west along the BNSF mainline near Vancouver, Washington. On April 20, 2001, at Wishram, Washington, the train stops for servicing.*

ABOVE: The SP&S No. 700 passes under the signal bridge entering the tri-cities area of Kennewick, Pasco, and Richland, Washington on April 20, 2001.

water supplied by the Columbia. Signs in Spanish and the colors of the Mexican flag on businesses reflect the heritage of the many workers who today make their way to the Inland Empire to harvest the carrots, beets, and potatoes of this fertile region.

In 1805, after an arduous crossing of the Rocky Mountains, Lewis and Clark passed this way descending the Columbia River. The Corps of Discovery in that year entered the mighty Columbia near the present-day town of Kennewick. At that time, the members of this first expedition in search of the Pacific could not have imagined the changes that would occur to this region over the next 200 years.

ABOVE: The restored Northern from Portland, rounds a curve along the Columbia River on April 20, 2001, on the run called the Homecoming Excursion.

On the second day of her homecoming tour, No. 700 encountered mechanical problems. Just west of Beatrice, Washington, unexpected wear on valve rings brought No. 700 to a standstill. The passenger excursion required a BNSF diesel helper from a nearby freight train. The addition of a diesel behind the steam locomotive did not seem to mar the enthusiasm of hundreds of Spokane citizens who turned out to welcome No. 700 home.

The crew made temporary repairs in Spokane, and the steamer returned to Pasco on April 22. Unfortunately, on Monday, April 23, two head-end cars derailed as the train prepared for departure. Passengers traveled by bus back down the Columbia River to Portland, and the steamer and her train deadheaded to Brooklyn Yard.

The crew of No. 700 took the disappointments of mechanical problems in stride, and repairs commenced back in Portland. In 2003, No. 700 pulled a flawless trip for Montana Rail Tours over former Northern Pacific track from Sandpoint, Idaho, to Billings, Montana, and back.

About the future of No. 700 and the PRSA, Jim Vanderbeck emphasized, "We are trying to keep this locomotive as a local asset. We feel that for years there will be a niche for us here."

RIGHT: The restored 4-8-4, SP&S No. 700, is caught racing along the Columbia River on April 20, 2001, en route to Pasco, Washington, on the first day's journey of the Homecoming Excursion.

700

Above: Restored Mountain type Frisco No. 1522, waits for the cameras to roll in Weston, Missouri, on May 5, 1995, during the filming of the HBO movie Truman.

16 Frisco No. 1522: Movie Star

President Harry Truman appeared hopelessly behind in his campaign for re-election in 1948. However, the president had a strategy. He planned one great trip across the country by train. The Democrat from Missouri took his campaign to the people.

Truman's campaign by railroad covered 21,928 miles in 33 days. He crossed the country twice. His efforts worked; in the end, Truman beat challenger, Thomas E. Dewey. One of the most famous photographs of that year was taken with President Truman on the rear of his observation coach, *Ferdinand Magellan*, holding up the *Chicago Tribune* headline mistakenly proclaiming "Dewey Defeats Truman."

When HBO Pictures decided to make a film portraying the story of the 33rd president, it took its production company to the Kansas City, Missouri, area near Truman's home in Lamar and his presidential library in Independence, Missouri. Casting agents chose actor Gary Sinese to star as Harry Truman. The company also found a nearby steam locomotive and train to star as the train that carried Truman on his Whistle-Stop Campaign. In early May 1995, Frisco No. 1522 and a small number of coaches from the Museum of Transportation in St. Louis rumbled across the state to western Missouri for their supporting role in this film.

Right: Frisco No. 1522 steams toward St. Joseph, Missouri, on the Burlington Northern mainline on May 4, 1995, during the filming of Truman.

RAIL ROAD
CROSSIN
2
TRACKS
1522
FRISCO

RIGHT: Early in the morning of May 5, 1995, Frisco No. 1522 steams south from St. Joseph, Missouri, to the small town of Weston, Missouri to begin the day's filming of the HBO movie Truman.

ABOVE: The feed mill of the Weston Elevator Company sits adjacent to the tracks of the Burlington Northern mainline in Weston, Missouri, where filming for Truman *occurred on May 4–5, 1995.*

Wayne Schmidt, director of the Museum of Transportation, supervised preparations of the train for the journey to Kansas City. Since the *Ferdinand Magellan*, the presidential coach President Truman actually used in his Whistle-Stop Campaign, was in the Gold Coast Museum in Miami, Florida, Schmidt chose the observation coach *Barrett Station* as a stand-in.

He recounted how museum workers spent weeks making the *Barrett Station* coach worthy for the supporting role in *Truman*. "The car was in pretty good shape," Schmidt said. "We did some work on the brakes and equalizer springs to allow it to run at normal track speeds."

Workers used photographs of the real *Ferdinand Magellan* from the Truman Library in Independence to transform the coach into its screen-ready appearance. Technicians mounted loudspeakers on the roof of the coach, and a replica of the presidential seal attached to the rear railing of the car to added authenticity. The U.S. government loaned the presidential seal to the movie company and required return of this emblem of the presidency when filming was completed.

The train traveling to Kansas City included the museum's commuter coach, No. 7200; the private car, *Choteau Club*; and No. 1522's two tool cars. The movie train was short compared to Truman's in 1948. The President's train left Washington, D.C., in 1948 with 17 coaches.

1522

PULLMAN

The St. Louis Steam Train Association (SLSTA) had No. 1522 in prime condition for her starring role in *Truman*. Baldwin Locomotive Company originally built the steam engine, a Mountain type 2-8-2, in 1926 for the Frisco—officially known as St. Louis–San Francisco Railroad. The steam engine pulled numerous passenger trains across the Southwest during her active career. In the 1950s, as diesels replaced steam on American railroads, the Frisco gave No. 1522 to the Museum of Transportation. A group of volunteers organized into the SLSTA and restored the locomotive in 1988.

On May 4, 1995, No. 1522 pulled onto a siding south of Weston, Missouri, and provided the set for the opening scenes of the movie. Catering tents, dressing room trailers, and large lights surrounded the steam train. Extras dressed as conductors, porters, and secret service agents milled around the locomotive awaiting their cues. Later in the afternoon, No. 1522 took the mainline and did several run-bys with Gary Sinese as Truman on board the observation coach. Lighting attached to the outside of the car provided illumination into the coach interior to simulate sunlight. Filming continued on into the evening before No. 1522 traveled north to St. Joseph, Missouri, for the night and the production team moved their equipment to Weston for the next day's shooting.

LEFT: Vintage cars and extras in period clothing wait for filming to begin on the railroad set of Truman. *The coach* Barrett Station *has been dressed to appear as President Harry Truman's campaign coach,* Ferdinand Magellan, *in his 1948 Whistle-Stop Campaign.*

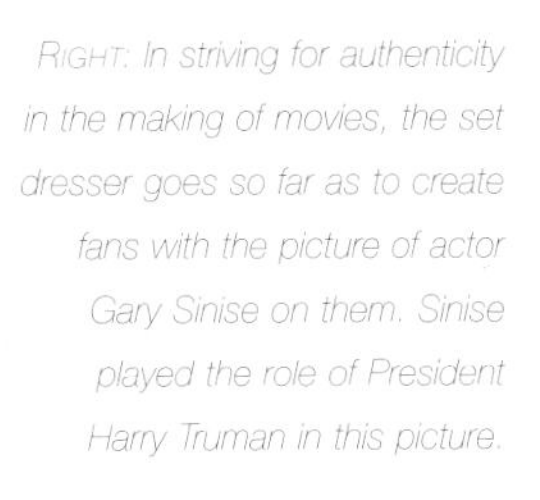

RIGHT: In striving for authenticity in the making of movies, the set dresser goes so far as to create fans with the picture of actor Gary Sinise on them. Sinise played the role of President Harry Truman in this picture.

1522

LEFT: Frisco No. 1522 took a break from her movie appearance to carry a trainload of U. S. war veterans from Kansas City to Lenexa, Kansas, as part of a patriotic celebration on May 6, 1995. The veterans' train passes a grain silo in on the outskirts of Lenexa.

LEFT: The observation coach, Barrett Station, *from the Museum of Transportation in St. Louis, has been wired for sound as external rigging is applied during the filming of the movie,* Truman, *outside of Weston, Missouri, on May 4, 1995.*

Early morning sunlight washed over the old Burlington depot in Weston as Frisco No. 1522 stormed into town at 50 miles per hour and ran past HBO cameras for the first shot of the day on May 5, 1995.

The film company's makeup artists prepared a crowd of extras dressed in period clothing as the train was backed onto the siding next to the Weston depot. The Pleasant Hill High School band waited on the freight dock of the station practicing two songs for their scenes in the movie, "Hail to the Chief" and "The Missouri Waltz." President Harry Truman hated "The Missouri Waltz," but bands played it for him anyway on his Whistle-Stop Campaign in 1948.

Vintage cars and red, white, and blue bunting on the ends of the depot completed the scene and actually gave the illusion of stepping back into the 1940s. For the actors, movie-making on location involved a lot of waiting while cameras and lighting were set properly for the next shot. During breaks, extras lounged around the depot or sat on their suitcases or street curbs while waiting for their cues in the next scene.

Filming sometimes had to stop to allow the rumble of a passing Burlington Northern freight train on the nearby mainline. The people of Weston took these delays in stride and seemed to enjoy their part in recreating scenes from the past for a movie about a president from their home state.

3985
UP
4-6-6-4-4-69 21 32 21 407 MB

17 Cheyenne to Laramie: Sherman Hill Stands Between

In the litany of special places in a railfan's geography, several choices come to mind. Chama, New Mexico, and Cumbres Pass always conjure up images of men and machines fighting blizzards on the narrow gauge of southern Colorado and northern New Mexico. Altoona, Pennsylvania, and nearby Horseshoe Curve is remembered as the fabled stomping ground of Pennsylvania K-4s on a four-track mainline, once so important to the nation that it was guarded during World War II.

The list could go on and include many favorites. But one place in the West still remains such an important site in American railroading that the entire city and region should be eligible for national park status. That city is Cheyenne, Wyoming. Included in that unique part of railroad geography is Sherman Hill, which lies west of Cheyenne in the mountains along the mainline of the Union Pacific Railroad.

The UP first reached the site that would become Cheyenne in the fall of 1867. The work crews conquered Sherman Hill by the spring of 1868. When the rails finally stretched to the town site of Laramie, Wyoming, new enthusiasm energized the builders. The railroad had finally reached the halfway point to Ogden, Utah, and completion of the country's first transcontinental railroad. The spirit must have been similar to entering orbit around the moon for the first time before a lunar landing in the 1960s.

Below: UP No. 3985 charges around a curve on Sherman Hill, Wyoming, in one of many run-bys made for the Rocky Mountain Railroad Club excursion on June 17, 1989.

Left: The engine crews of the Union Pacific Challenger No. 3985 takes a break before departing Cheyenne for Laramie on June 17, 1989.

Above: UP Challenger No. 3985 has been tucked away inside the roundhouse in Cheyenne, Wyoming.

Cheyenne grew from a cluster of tents on the prairie in 1867, to a town of nearly 5,000 residents by 1875. Cattle ranching came to the rural towns of southern Wyoming and became a major industry. But the main business for Cheyenne remained the Union Pacific Railroad.

The first UP depot in Cheyenne was a small wood-frame building. In 1868, the railroad laid the cornerstone for a new magnificent granite station designed by the architectural firm of Van Brunt and Howe of Kansas City and Boston. The architects employed design features in the Richardson Romanesque style with repeating rock arches over the windows and doorways. The impressive tower of the UP depot still serves as a landmark for both rail and highway travelers.

Over the years, trains with names like *City of San Francisco* and *City of Los Angeles* have paused in Cheyenne on their way between the West Coast and Omaha, Nebraska. The depot saw its last regularly scheduled passenger train in 1983, when the *Rio Grande Zephyr*. The latter was discontinued in Colorado to allow Amtrak's *San Francisco Zephyr* to move from the UP mainline down to Denver, Colorado, and the Rio Grande's Moffat Tunnel route to Salt Lake City. In that year regular passenger service disappeared for good on the Overland Route through Cheyenne.

The Cheyenne depot has recently been renovated and is today a visitor center and museum. The UP depot received permanent protection in 1973 when it was added to the National Register of Historic Places.

ABOVE: UP Challenger No. 3985 makes one of 14 run-bys in a day's excursion trip from Cheyenne to Laramie on June 17, 1989. The austere landscape of Sherman Hill is covered in steam and smoke of the Challenger locomotive.

RIGHT: UP No. 3985 storms under a signal bridge on the UP mainline at Tie Siding. The train is en route back to Cheyenne after an excursion to Laramie in 1983.

X3985
UNION
PACIFIC
3985
3985

ABOVE: The shops of the Union Pacific Railroad in Cheyenne, Wyoming, provide protection for the vintage equipment the railroad uses for excursions, publicity, and business trains. UP Challenger No 3985 is in steam in the spring of 1983.

LEFT: Just outside of Cheyenne, UP No. 3985 awaits an excursion train from Denver, Colorado, in 1983. The Challenger will haul the train on to Laramie, Wyoming.

Through the years of World War II and into the 1950s, Cheyenne, Wyoming, came to be a mecca for enthusiasts of big-time steam railroading. Heavy pulling power was needed to master the grades of Sherman Hill in the Laramie Mountains west of Cheyenne. The Union Pacific employed large locomotives like Big Boy 4-8-8-4s and Challenger 4-6-6-4s for heavy tonnage over the hill.

The Union Pacific Railroad appreciated its place in history as the nation's first transcontinental railroad. As steam disappeared from service on the UP in the late 1950s, several locomotives were preserved in the Cheyenne shops and would later enter active excursion service.

UP No. 844 was the last steam engine built for the road in 1944 by the American Locomotive Company in Schenectady, New York. UP No. 844 carries the title of being the only steam locomotive never out of service since she was built.

A second active steam engine occupies the roundhouse in Cheyenne. UP No. 3985 is a Challenger type, 4-6-6-4, retired in 1962 from active freight service and returned to excursion travel in 1981. Together these two steam locomotives have roamed the UP system across the country, bringing delight to the hearts of a new generation of railfans. Some of their most memorable trips have been over Sherman Hill from Cheyenne to Laramie.

ABOVE: UP No. 8444 shares the roundhouse in Cheyenne, Wyoming, with No. 3985.

In moving trains over the Laramie Mountains west of Cheyenne, the Union Pacific Railroad has always fought blizzards and extreme weather in this landform resembling a lunar surface. Large wind-sculpted mountains and precariously balanced boulders guard the way to the summit of Sherman Pass. These superb geographical features have attracted photographers to this major league show of railroading for the past 130 years. Anyone who plants a tripod and camera on the windswept mountains along the UP mainline in Wyoming follows in the footsteps of scores of celebrated photographers such as Andrew Russell, Richard Kindig, Jim Shaughnessy, and Jim Ehernberger.

It would be a close call between two complementary organizations to determine which has sponsored the most rail excursions over Sherman Hill. The Rocky Mountain Railroad Club and the Intermountain Chapter of the National Railway Historical Society, both of Denver, ran annual trips in years past. The journey over Sherman Hill to Laramie always ranked high as a favorite. Travelers unloaded from the train in the grassy lawn next to the Laramie depot and explored the restaurants of the community before the afternoon departure back to Cheyenne. One trip by the Rocky Mountain Railroad Club boasted no few than 14 run-bys for railfan photographers.

RIGHT: UP Challenger No. 3985 pulls through an elegant curve on the UP mainline over Sherman Hill, Wyoming, during a Rocky Mountain Railroad Club excursion on June 17, 1989.

X3985
X3985
UNION PACIFIC
3985
3985

18 Union Pacific No. 3985 Along the Route of the Old Missouri Pacific

In January 2004, Union Pacific's Challenger No. 3985 traveled south from the high plains of Cheyenne, Wyoming, to the Super Bowl in Houston, Texas. The locomotive's return trip north in February brought her through Arkansas. While most railfan photographers enjoy the chance to photograph a vintage steam locomotive on a historic home track, the opportunity to photograph any steam engine on modern mainline rail is a chance not to be missed.

Leaving North Little Rock the morning of February 6, 2004, in a cloud of steam and smoke, the Challenger had the dome of the state capitol building across the river in Little Rock as a backdrop. The locomotive traveled along the former Missouri Pacific Railroad line that started life as the Little Rock and Fort Smith LR&FS Railroad in the years after the Civil War.

Arriving in Conway, Arkansas, the train paused to drop off passengers before continuing westbound. Conway, the county seat of Faulkner County, owed its existence to the LR&FS Railroad. Colonel A. P. Robinson came to what would become Conway after the Civil War. He was chief engineer for the fledgling railroad building between the capital city and Fort Smith, Arkansas, on the border of Indian Territory in Oklahoma. Robinson gave part of his land to the city for a depot site and called it Conway.

By 1906, the FS&LR had gone through several reorganizations and became part of the St. Louis, Iron Mountain & Southern Railway. By 1917 the Iron Mountain merged into the modern Missouri Pacific (MP) system.

RIGHT: Union Pacific Challenger No. 3985 is far from her home in Cheyenne, Wyoming, as she bursts through the Conway tunnel on the former Missouri Pacific line.

ABOVE: UP No 3985 arrives at the restored Missouri Pacific depot in Russellville, Arkansas, on February 6, 2004, and is greeted by hundreds of citizens eager for a look at this traveler from Cheyenne, Wyoming.

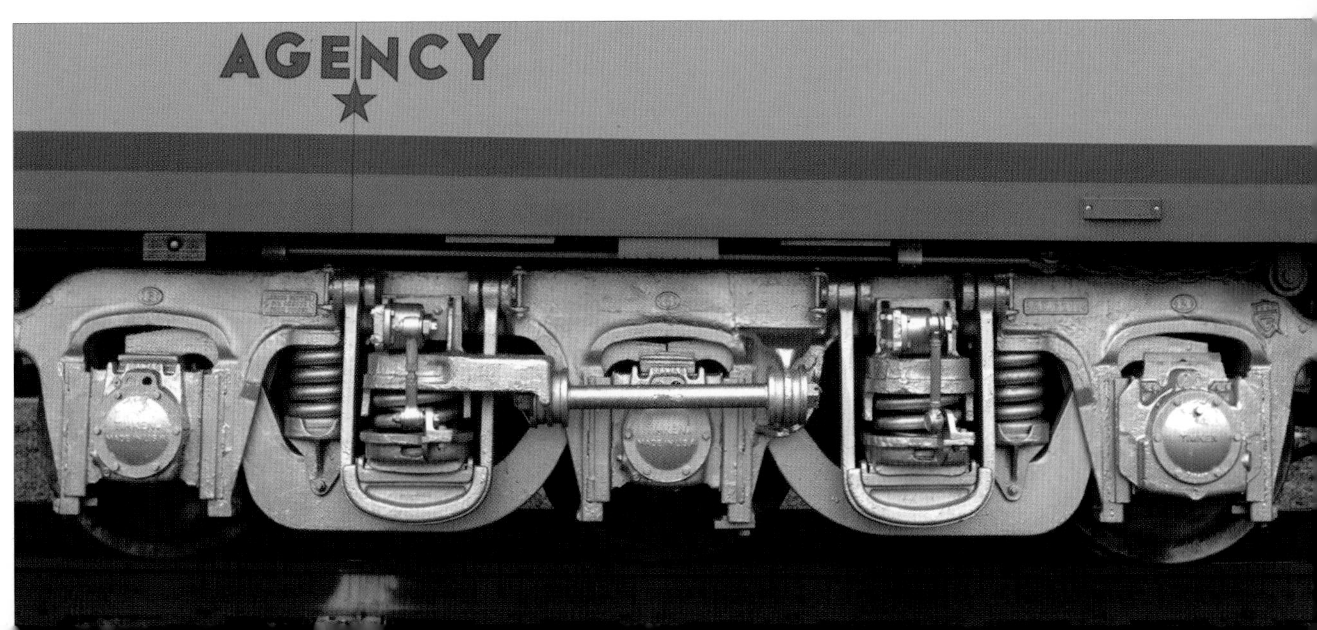

RIGHT: The running gear of the baggage coach on the Union Pacific train in Russellville, Arkansas, on February 6, 2004, provides an abstract study of the equipment of railroading.

The Conway tunnel on the old Missouri Pacific was situated directly beneath the roadbed of Route 64 east of Conway. Drivers on the highway that day in February may have been surprised to see a steam locomotive pop out of the tunnel in a veil of smoke and steam. The Conway tunnel is the only one along this stretch of the former Iron Mountain section of the MP.

A palpable rush of wind greeted the photographers gathered at the west end of the tunnel to record the passing of the Challenger. The tunnel took 15 minutes to clear the smoke, but few photographers were there to witness the event. Most were on the road in hot pursuit of the steam engine, looking for their next setup for photography or video.

Finding depots on any railroad is unusual in the new millennium, but several survived along the UP in Arkansas. At Morrilton, schools brought students to the tracks to watch the passage of the Challenger from Cheyenne. The depot in this small town served as a focus for the gathered onlookers.

In Russellville, Arkansas, No. 3985 pulled into the newly restored depot for a service stop. A local fire truck arrived to pump water into the tender. Once again, schools in the city allowed students to come to the depot to see a sight that had once been familiar to their grandparents.

RIGHT: Winter temperatures provide a show of drama and steam as UP No. 3985 is readied for the trip from North Little Rock to Van Buren on February 6, 2004.

AMERICAN

Left: Early on the morning of February 7, 2004, in Van Buren, Arkansas, a Pierce-Arrow has joined UP No. 3985 on the ready track as the Super Bowl Special heads back west toward Coffeyville, Kansas, and eventually Cheyenne, Wyoming.

West of Russellville, the UP Challenger crossed Big Piney Creek and Illinois Bayou, where the two creeks empted into Lake Dardanelle on the Arkansas River. The train was often enveloped in a cloud of steam as the air temperature dropped steadily throughout the day. The forecast included snow flurries for the late afternoon.

At Coal Hill, the *Super Bowl Special* ran past sites where coal had once been loaded onto Missouri Pacific trains for export to other parts of the country. At Clarksville, Arkansas, the UP line passed the cutoff into this county seat town for Johnson County. In the days of passenger service, MP trains pulled into Clarksville on a spur. Today that spur is a hiking/biking trail.

A sizable crowd turned out at the town of Ozark, Arkansas, for the visit of No. 3985. This town of several thousand is situated adjacent to the shores of the Arkansas River. The old stone MP depot is preserved today and is on the National Register of Historic Places. The Highway 23 bridge in town provided an elevated platform for photographers as the train steamed around the curve through town and under the bridge. West of Ozark, the train entered a rock cut right at water's edge on the Arkansas River. The weather grew colder and snowflakes began to fall as the train left Ozark and steamed toward Van Buren.

Above: A crewmember of UP No. 3985 waits in the cab for the morning departure from North Little Rock, Arkansas, on February 6, 2004.

UNION
PACIFIC
3985
3985

Above: UP No. 3985 sits for a portrait in North Little Rock, Arkansas, on February 2004.

Above Right: A driving wheel of UP Challenger No. 3985 provides a study in abstraction on February 6, 2004, in Van Buren, Arkansas.

Left: Union Pacific No. 3985 emerges from the mists and steam in a ghostly fashion at North Little Rock on the morning of February 6, 2004. The locomotive and train traveled along the former Missouri Pacific line to Van Buren, Arkansas.

The MP depot at Alma, Arkansas was long gone, but fruit and produce sheds still lined the tracks. The UP Challenger and her train paused in this small town known for its spinach production and canning company. A number of passengers left the train in Alma before No. 3985 made the final leg of the day's journey to Van Buren, Arkansas, for an overnight stop.

By the time the UP train reached Van Buren, the temperature had dropped into the upper 20s and snow was falling. In spite of the cold temperatures, hundreds of people came down to the UP tracks in Van Buren to welcome the Challenger and her train.

On Saturday morning, February 7, 2004, it seemed that everyone who had been present the evening before returned to the Van Buren railroad yard to see the Challenger leave town for Coffeyville, Kansas. The atmosphere again resembled a circus. Small children enjoyed the shoulders of their dads as people excitedly surrounded the locomotive. When the UP steam might return was not known, and no one wanted to miss one last chance to see the visitor from Cheyenne.

19 Carl Franz and the "Photo Special": Mountain Thunder on the Western Maryland

When the human genome is fully mapped, there will have to be a sequence that identifies railfans. People who love trains seem to have similar stories. At a formative time—usually around the age of 10 years in those with this genetic predisposition—something is triggered that sets in motion a life of fascination with trains and railroading.

For Carl Franz, this triggering mechanism was a childhood spent near the Baltimore & Ohio (B&O) Railroad in Cleveland, Ohio. "The B&O tracks were 500 feet from my house," said Franz, remembering how his life with trains began. "As a 10-year-old boy, I would ride my bike down to the Bagley Road Lumber Yard when I heard the 'local' blowing for the crossing. The fireman, Bill Weinbrewer, would invite me up into the cab while they dropped off loads of lumber and picked up the empties," said Franz about his lifelong love of trains.

During the summers of his early teen years, Carl rode the cabs of B&O 2-8-0s at least once a week. Franz was also active in the local chapter of the National Railway Historical Society (NRHS), for which he assisted with rail excursions.

RIGHT: The WMSRR No. 734 steams past a house in Frostburg, Maryland, on October 18, 1999.

RIGHT: Western Maryland Scenic Railroad No. 734 enters a curve on one of many photo run-bys in the trip sponsored by Carl Franz in 1999.

734
WESTERN MARYLAND

Right: The Western Maryland Scenic No. 734 is at the entrance to Brush Tunnel waiting for rail photographers to get images of the train.

Above: WMSRR No. 734, a 2-8-0, stops at the entrance to Brush Tunnel to allow photographer to enter and set shots from the tunnel interior.

Carl took photography courses in high school to improve his photos of trains. This interest led to a Bachelor of Science degree in photographic engineering from the Rochester Institute of Technology of New York in 1965. He then moved to Washington, D.C., and took a job designing photo-imaging systems at a U.S. Navy research laboratory.

His work in optical instrumentation took him around the world. One of his many projects involved the design of a long-range imaging system for collecting military intelligence data; it had an effective equivalent of a 30,000-millimeter on a 35-millimeter camera. "The system could clearly resolve the coarse grains of sand in mortar on a church steeple a mile away," he said with a laugh.

Franz became active in the Washington and Potomac Chapters of the NRHS in D.C. area soon after his arrival. He helped organize trips on the Southern Railway during this time. "We had a hard time getting people to be car hosts," said Carl. "Photo run-bys were often limited to one or two on some excursions," he added. People said they would rather chase the train and set up more shots. Franz decided he would organize a charter trip and give people what they wanted.

Friends thought he wouldn't fill all 25 spots on his first steam excursion in 1975 on the Cass Scenic Railroad. "We sold out very quickly," he said. The Carl Franz steam "Photo Specials" were off and running.

ABOVE: WMSRR No. 734 is positioned at the depot in Cumberland, Maryland, for day's trip to Frostburg on a Carl Franz Photo Special.

LEFT: Photographer's special freight has returned to Cumberland, Maryland, late in the evening of October 18, 1999, after a day of shooting which included over 40 run-bys.

At last count, Carl Franz has organized over 120 steam trips across the country on seven different railroads. Thirty-five of those Photo Specials have been on the Western Maryland Scenic Railroad (WMSR). At first he just wanted to improve his own rail photography. As time passed, Franz came to realize that his real joy was in running trips so other people could have fun.

A typical Carl Franz trip on the WMSR begins at the red-brick depot in downtown Cumberland, Maryland. The restored depot also houses the visitor center for the western terminus of the Chesapeake & Ohio Canal Historical Park, part of the U.S. National Park Service.

As WMSR No. 734 thunders up the grade to Frostburg, Maryland, on the 34-mile roundtrip, stops are usually made at Helmstetter's Curve, a horseshoe bend in the line encompassing the scenic farm for which the site is named. Brush Tunnel and a number of trestles add interest to photo stops before the train arrives at Frostburg.

"The Western Maryland is one of the few railroads in the east where you can see the locomotive working hard to climb a grade. You get a feel for what steam railroading was really about," Carl said.

OPPOSITE PAGE: "The Western Maryland is one of the few railroads in the east where you can see the locomotive working hard to climb a grade. You get a feel for what steam railroading was really about," says Carl Franz.

BELOW: WMSRR No. 734 offers photographers a chance to enjoy eastern railroading in the Allegheny Mountains splashed with the colors of autumn.

The Western Maryland Scenic Railroad remains one of the East Coast's prime sites of steam preservation. However, the railroad has not been above political entanglements. The railroad received a subsidy for most of its existence from the state of Maryland that is now scheduled to cease. The non-profit operators think they can still run the railroad with a small increase in ticket costs.

The railroad will also have to share Brush Tunnel with a hiking/biking trail in the near future. The railroad and leaders of the Allegheny Highlands Trail of Maryland will have to work out safety issues concerning tunnel use. In the end, both Cumberland and the WMSR may benefit from the hiking trail. The railroad already plans to add a baggage car to carry bicycles up the mountain grade to Frostburg and allow cyclists to ride back downhill to Cumberland.

As Carl Franz prepared for a new season of excursions, he thought back to his boyhood days along the B&O: "If it hadn't been for that B&O fireman who gave me cab rides, I might have grown up to be normal."

734
734
734
WESTERN MARYLAND

20 Norfolk & Western No. 611: In Search of O. Winston Link

In the first decade of the twenty-first century, a quarter-million dollars buys a nicely equipped Cessna 182 turbocharged airplane. For the same price tag in 1950, the Norfolk & Western Railroad built its last streamlined J-Class steam locomotive, No. 611.

A Cessna 182 cruises at 145 knots and carries four people. No. 611 ran at 70 miles per hour and pulled coaches for up to 600 people, whom rode comfortably between Norfolk, Virginia, and points west all the way to Cincinnati, Ohio. The Cessna is a nice airplane, but the N&W got a great deal and its handiwork still endures.

Locomotive No. 611 entered service on May 1950. After just nine years, this elegant bullet-nosed streamlined locomotive pulled her last revenue train. Running roundtrip between Roanoke, Virginia, and Bluefield, West Virginia, on October 24, 1959, No. 611 outlasted many of her steam engine counterparts on other railroads at a time when diesels were taking over the rails and Americans were taking to the highways in fancy cars with tailfins and chrome.

This J-class locomotive, a Northern type, produces the equivalent power of 5,200 horses with a tractive effort of 80,000 pounds. Her driving wheels stand 70 inches high and she weighs in at a dainty 494,000 pounds.

After retirement, the N&W Railway donated No. 611 to the Virginia Museum of Transportation in Roanoke. Railfans rejoiced in 1981 when No. 611 was restored and returned to service as a workhorse of the N&W steam program.

Right: N&W No. 611 has just pulled an excursion from St. Louis to Moberly, Missouri.

Above: N&W No. 611 waits in Moberly, Missouri, after the trip from St. Louis in summer 1983.

Right: N&W No. 611 waits to board passengers for the return trip to St. Louis in mid-1983.

Anticipating the schedules of the Norfolk Southern steam program during the 1980s and early 1990s became one of the joys of summer. About the time of baseball's spring training, excursion trips were published in *Trains* and *Railfan and Railroad* magazines. Whether it was No. 611 or her bigger brother, N&W No. 1218, pulling the excursion, vacation plans were often forged around the routes and times of these apparitions from the past.

N&W No. 611 made several trips up and down the Shenandoah line in Virginia. She traveled out west to near Cincinnati and even across the Mississippi River into Missouri several times. Excursions from either Kansas City or St. Louis often turned around on the former Wabash line in the central Missouri town of Moberly.

Even though the route across the Show-Me state resembled the Blue Ridge of Virginia, home territory for the N&W steamer, the topography of Missouri was not the route of the old *Powhatan Arrow*, *Pocahontas*, or the *Cavalier*. Most railroad photographers hoped someday to photograph the two N&W steamers on their home ground back east.

RIGHT: Norfolk & Western Northern No. 611 crosses the Missouri River bridge leaving St. Louis for a trip to Moberly in the summer of 1983.

ERN
611

BELOW: The clock of the restored Cincinnati Union Terminal in Cincinnati, Ohio, says 8:30 a.m. on a September morning in 2004. N&W train used Cincinnati Union Terminal as its western terminus.

O Winston Link photographed the last days of steam railroading on the N&W Railway in 1957. Readers of railroad magazines appreciated his work, but his audience remained limited for 30 years. In the 1980s, Link pulled his negatives out of storage and skyrocketed to fame in the fine art world of galleries and museums across America. He became the Ansel Adams of railroad photography. Prices rose into the thousands of dollars for his luminous 16x20-inch exhibition prints which captured the feel of a time long past.

Link's prints were technically perfect, and most were shot after dark using gigantic banks of Sylvania No. 2 flashbulbs. Sometimes, Link could use several thousand feet of wiring and the whole length of one day to set the lighting for one passing train in the night.

ABOVE: "To Trains" reminds travelers in Cincinnati Union Terminal of the days when N&W trains, such as the Pocahontas, called on the city with the Art Deco station.

Many of Link's prints featured J-Class streamlined Northerns. No. 611 even made an appearance in several shots by Link during those two years of shooting. What really made Link's photographs stand out was the placement of people, who lived and worked along the line, within the frames. Link's photographs of the N&W left us an invaluable archive of life in the Shenandoah Valley during the middle of the twentieth century.

Visitors today follow in Link's footsteps and visit those places made famous by his black-and-white images. The gravity-fed gas pump is long gone at the old general store in Vesuvius, Virgina, where Link made one of his most famous photographs, "When the Electricity Fails." But the memory of fast trains in the night lingers.

ABOVE: Travelers can retrace the steps of O. Winston Link along the Shenandoah lines of the old Norfolk & Western in Virginia and visit sites such as the general store where the famous photograph, "When the Electricity Fails" was made. This photograph by Link showed a young couple in a convertible as an elderly man pumped gasoline into their car and a N&W locomotive passed behind them on the nearby mainline.

NORFOLK AND WESTERN

Left: N&W No. 611 passes under a highway overpass headed to Moberly in the summer of 1983 on an excursion from St. Louis.

The Claytor brothers of Virginia had railroading in their blood. Robert became president of the Norfolk & Western Railway in 1981. His older brother, W. Graham Claytor Jr., trained in law and served as president of the Southern Railway for many years. At the time Robert took the reins of the N&W, Graham was in charge of the Southern Railway's successful steam program that had sent passenger excursions across the southeastern United States since the 1960s.

Robert Claytor wanted a return of steam to the N&W. He arranged for the lease of No. 611 from the Virginia Transportation Museum and sent the locomotive down to the Southern Railway's steam shops in Birmingham, Alabama. During No. 611's overhaul, N&W and Southern merged into the Norfolk Southern system in 1982. N&W No. 611 soon hit the rails again running over the new Norfolk Southern.

The Claytor brothers often found their ways into the cabs and at the controls of No. 611, as well as No. 1218. Graham went on to become Secretary of the Navy and chairman of Amtrak in the 1980s. Robert continued running steam trains across the south.

A switching derailment in September 1994 damaged several coaches in Lynchburg, Virginia. Rising insurance costs and the expense of repairing the damaged equipment doomed the NS steam service. No. 611 made her final trip back to her birthplace of Roanoke on December 7, 1994, and returned to the museum where her excursion career began.

Above: A restored Kansas City Southern observation coach brings up the rear of an excursion train on the Norfolk Southern powered by N&W No. 611. The train has returned to Kansas City, Missouri, on June 23, 1985.

18
18

21 The Engineer Was a Woman: Adventure on the Grand Canyon Railway

Glenn Miller music filtered down from the public address system as Marty Fischer coupled her steam locomotive onto the waiting train in Williams, Arizona. Dropping down to the platform of the Grand Canyon Railway's former Santa Fe Railroad depot, Fischer greeted fellow employees as they prepared the train for the morning run to the Grand Canyon's south rim.

Only a few passengers seemed surprised that this slim young woman in denim overalls and work shirt was their engineer for the day. The year was 1990, not 1910 when the only job for a woman might have been serving tables in the Fred Harvey Hotel adjacent to the Williams depot.

When asked if Fischer thought she was probably the only female steam engineer in the country, she replied, "I'm not sure. There may be two three more, but I just don't know them." She continued, "There was a woman fireman on the Durango–Silverton line when it was still the Rio Grande."

Marty reflected upon her unique status as a female railroader. In Chama, New Mexico, where she formerly worked, an elderly retired steam engineer was inspecting the narrow gauge engine about to depart. He looked up in the cab, saw that Marty was a woman running the locomotive, and walked away muttering, "Isn't anything sacred anymore?"

Fischer and her husband, Russ, moved from the narrow gauge country of Chama in the summer of 1989. Russ acquired the position of foreman of locomotives, as well as an engineer for the Grand Canyon Railway.

BELOW: The Grand Canyon Railway passenger train has arrived at the South Rim of the Grand Canyon in March 1990. Passengers have left the train to explore the beauties of the Canyon and have lunch in the elegant El Tovar Hotel on its lip. The hotel was built by the Santa Fe Railroad in the early 1900s and lured travelers from the east.

LEFT: Work crews at the Grand Canyon Railway in Williams, Arizona, examine No. 18 before the day's trip to the South Rim of the Grand Canyon. The date is March 1990.

Born and raised in the mountains of central New Mexico, Marty Fischer went to work on the snack car of the Cumbres & Toltec Scenic Railroad's narrow gauge steam train after finishing high school in 1977. She met Russ Fischer on the railroad, and they were married in 1981.

"I was interested in the locomotives—shoveling coal and learning how to run an engine," said Marty about her path to becoming an engineer, "but I didn't think I was physically capable at first." She continued, "You have to shovel 2 tons of coal an hour."

With the encouragement of Russ, she went to work convincing the C&TSRR management that she should be given the chance to run a steam locomotive. After a couple of years, she received her request, started as brakeman, and then became a fireman. "I was pleasantly surprised to learn that I could do the job," she said

During their years in Chama, the couple's employment was seasonal. Russ was one of only a handful of employees retrained through the winter by the C&TS. Marty's work always ended in October when snows came to the mountains around Chama and the tourist line closed for the winter

In 1988 Arizona businessman, Max Biegert, proposed reopening the Santa Fe's former Grand Canyon line. The 65-mile-long spur from Williams to the South Rim of the Grand Canyon originally opened in 1901. The last Santa Fe passenger train passed over the route in 1965.

RIGHT: The restored Santa Fe depot and Fred Harvey Hotel in Williams, Arizona, serves as a gateway to adventure to one of America's most spectacular national parks, the Grand Canyon. The railroad offers travelers a chance to arrive at the South Rim in style.

GR
2133

Max Biegert's plan to reopen the rail line to the Grand Canyon gave the Fischers a chance for both to have fulltime employment in the rather unique industry of modern-day steam railroading. "It was a good opportunity for us to get in on the ground floor and work for a big company," said Marty.

As Beigert began renovation of the former Santa Fe depot and the adjoining Fray Marcos Hotel (part of Fred Harvey Hotels), he also began assembling the skilled people needed to renovate and operate the steam-era equipment he purchased for the new tourist railroad.

The Fischers had never been to the Grand Canyon before their move to Williams. "We knew they were serious about restoring the line when we saw the pine trees gone from the yard at the Canyon," said Russ, referring to the large trees that had grown between the rails in front of the depot at the South Rim.

During August 1989, Fischer and his coworkers, Robert Crossman and Greg Griffin, battled the clock while overhauling Consolidation No. 18 in time for the inaugural run from Williams to the South Rim on September 17. The Alco 2-8-0 was one of four locomotives of the same type purchased by the new Grand Canyon line. All had once hauled coal for the Lake Superior and Ishpeming Railroad in Michigan.

RIGHT: The Castaneda Hotel in Las Vegas, New Mexico, survives today as an example of the chain of Fred Harvey Hotels on the old Santa Fe Railroad that welcomed travelers from Chicago to Los Angeles, and especially at the South Rim of the Grand Canyon.

BELOW: Workers of the Grand Canyon Railway inspect locomotive No. 18 before the train leaves Williams, Arizona, for the South Rim of the Grand Canyon in March 1990.

RIGHT: Grand Canyon Railway No. 18 has returned from the day's run to the South Rim of the Grand Canyon in Arizona in March 1990. The train serves as an economical, environmentally friendly way to move passengers effectively from Williams into the national park.

ABOVE: No. 18 has just returned from the South Rim of the Grand Canyon to the station platform of the Railway's restored depot and Fred Harvey Hotel in March 1990.

RIGHT: The Williams, Arizona, depot is bathed in soft moonlight on a March evening in 1990.

The morning run to the Grand Canyon's South Rim began at 10 a.m. and took about two hours. As the route climbed toward the 7,000-foot southern lip of the Canyon, the broad-shouldered land became like gently rolling ocean swells.

When the train entered the rugged Coconino Wash, Marty Fischer could be seen leaning out the cab window coaxing No. 18 through the tight curves of the small canyon. "One of the greatest challenges in running a steam engine," she said, "is knowing the line. You really have to pay attention and stay on top of what your are doing."

The train eased into the log-style Santa Fe depot at the South Rim in a scene reminiscent of travel posters of the Indian Detours of the 1920s and 1930s. Passengers headed for the El Tovar Hotel that sits on the brink of the Canyon. Many passengers ate lunch in either the El Tovar or Bright Angel Lodge, both still run by the Fred Harvey Company, a name long associated with the old Santa Fe Railroad.

Later in the afternoon, Marty met Russ and some fellow railroaders who had driven up to the South Rim just to see the spectacle of their steam train leaving Grand Canyon Village. Mrs. Fischer didn't disappoint her friends and husband as she left Grand Canyon's South Rim in a fury of sound, steam, and smoke.

18

750
750
S&A

22 Movie Magic on the Arkansas & Missouri Railroad: The Making of *Biloxi Blues*

As long as Hollywood makes period movies about life in the 1940s, there remains a need for steam trains. During that decade in North America, trains seemed ubiquitous. They carried troops to war and materials to seaports for shipment to the front lines. Steam trains brought fresh milk and the day's mail to large cities and small towns across the country.

When Universal Studios decided to film the Neil Simon play, *Biloxi Blues*, it took its production company to Fort Smith, Arkansas, because of the nearby army compound of Fort Chaffee. The wooden barracks of the 72,000-acre camp, built during World War II, provided the stage for the telling of a young recruit's training in the early days of the war.

The movie opens with scenes of a steam troop train carrying the main character of the story, played by Matthew Broderick, to an army training post in Biloxi, Mississippi. Fort Chaffee, Arkansas, stood in for the camp in Biloxi, and Savannah & Atlanta (S&A) Railway No. 750 starred as the locomotive pulling the troop train.

S&A No. 750, owned by the Atlanta chapter of the National Railway Historical Society, traveled by rail to Arkansas just for the making of *Biloxi Blues*. The train consisted of former commuter coaches from the East Coast provided by a railroad equipment company.

LEFTN: Savannah & Atlanta (S&A) No. 750 steams before the camera of Universal Studios in the filming of the movie Biloxi Blues *in Chester, Arkansas, on June 25, 1987.*

RIGHT: On June 20, 1987, engine crew check the status of S&A No. 750 at Winslow, Arkansas, prior to the start of filming of railroad sequences for Biloxi Blues *on the Arkansas & Missouri (A&M) Railroad line south of Fayetteville, Arkansas.*

ABOVE: Steam cocks are opened at Winslow, Arkansas, on S&A No. 750 before the day of filming begins before the cameras of Universal Studios' Biloxi Blues.

RIGHT: Extras in uniform wait for their cues in the filming of Biloxi Blues *on the mainline of the A&M Railroad at Chester, Arkansas, on June 25, 1987.*

For eight days in early June 1987, S&A No. 750 paraded up and down the Arkansas & Missouri (A&M) Railroad's former Frisco line in the Ozark Mountains of western Arkansas. Universal Studios' cameras captured on film the footage necessary for the telling of *Biloxi Blues*, a story which begins and ends on a train.

Local Arkansas people provided the extras for the movie. They dressed in army uniforms, overalls, and gingham dresses. Antique automobiles and trucks paused on cue at railroad crossings in Chester and Van Buren as the No. 750 and her train ran passed by numerous times during each day of shooting.

Filming took place at several locations along the A&M Railroad. One whole day was spent at Chester, Arkansas. Slag from a local steel plant was placed on the road to cover the pavement and simulate dirt roads. A Cadillac sedan served as a pace vehicle. The film crew removed the lid from the rear trunk of a Cadillac and attached a Panavision camera. A stool in the trunk provided a chair for the camera operator. Improvisation seemed to be the order of the day.

Filming lasted all day and into the night when No. 750 provided classic shots as she ran past the Chester General store. Clips of this footage have been used in other later films such as *Seabiscuit* and *Fried Green Tomatoes*.

YOUR STEP
RFMX
3207

ATLANTA-NRHS
750

Above: S&A No. 750 steams past the restore Frisco depot in Van Buren, Arkansas, during the filming of Biloxi Blues.

Left: Universal Studios provides million-dollar lighting for still photographers visiting the set of Biloxi Blues during filming at Chester, Arkansas, on June 26, 1987.

Early on Saturday morning, June 20, 1987, the steam train, actors, and camera crew came together in Winslow, Arkansas, for a day of filming. The train rolled into place as an aerial film crew mounted a Panavision camera on a Bell Jet Ranger helicopter. Actors and extras took their places onboard the train.

There is only one tunnel on the Arkansas & Missouri Railroad that located just south of this small picturesque Ozark community in the rugged Ozark Mountains. Local railfan photographers hiked through the ½-mile-long tunnel and waited patiently. Finally, the train blasted out of the tunnel at the south entrance with the whistle blowing and steam cocks open. The helicopter hovered 10 feet above the tracks and lifted into the plume of smoke and steam as the train swept past beneath the landing skids of the chopper. The engine crew later related that this close encounter with the helicopter scared them to death. Smoke filled tunnel and rolled out of the mountain for 30 minutes.

On Monday morning, June 22, the train and camera crew gathered for filming in front of the former Frisco depot in Van Buren. Vintage cars and trucks parked around the depot provided the right air of authenticity when No. 750 rolled past the station.

750

LEFT: S&A No. 750 steams before the camera for the filming of Biloxi Blues at Chester, Arkansas on June 25, 1987.

ABOVE: A vintage pick up truck pulls to the grade crossing at Chester, Arkansas, on June 25, 1987, for the making of the major motion picture, Biloxi Blues.

Change is the one constant of life. Locomotive No. 750 returned to Atlanta after the movie and resumed her circumnavigation of Atlanta on the New Georgia Railroad. Eventually she was retired due to wear and tear beyond feasible economic restoration. The locomotive is now on static display at the Southeastern Railway Museum in Duluth, Georgia. Whether this elegant locomotive will ever steam again remains unknown. But No. 750 had her day in the spotlight in a classic movie about the 1940s.

The Ozark Mountains between Fort Smith and Fayetteville also changed significantly when Interstate 540 was built in the years following *Biloxi Blues*. The movie footage of this steam train along the A&M—especially the aerial material—will be of interest to historians and cultural geographers, as the booming northwest Arkansas economy changes the landscape in the rugged Ozarks Mountains.

Biloxi Blues came at the right time to stimulate a regional interest in passenger trains on the Arkansas & Missouri Railroad's old Frisco line. The future seems bright for excursion trains on the A&M, but change is the one constant of life. Perhaps a steam engine will appear again, like a ghost from the past, on its rails.

23 Go North for Steam: Canadian Pacific No. 2816

In 1998 Robert Ritchie, president and CEO of the Canadian Pacific Railway (CPR), wanted a steam engine. Fortunately, a preserved CPR Hudson existed at Steamtown National Historic Site in Scranton, Pennsylvania. CPR No. 2816 left her homeland of Canada in 1968 and traveled to Nelson Blount's *Steamtown USA* in North Walpole, New Hampshire. This museum collection, including No. 2816, eventually moved to Scranton after a short sojourn in Bellows Falls, Vermont.

Montreal Locomotive Works built Canadian Pacific No. 2816 in 1930. During her active career, No. 2816 moved crack CPR passenger trains over the middle provinces of Canada and then finished out pulling commuter trains in the Montreal area in the late 1950s as diesels invaded Canadian railroads.

When Ritchie learned of the Class H1b Hudson, he also decided that the man to restore No. 2816 should be Al Broadfoot, who was then running the British Colombia Rail steam shops in Vancouver, British Columbia.

RIGHT: Static steam tests are performed on Hudson No. 2816 on January 30, 2006, in CPR shops in Calgary, Alberta.

BELOW LEFT: The face of Canadian Pacific No. 2816 has been removed to allow access to the interior of the Hudson type locomotive in the CPR shops in Calgary, Alberta, in January 2006.

BELOW: The face of CPR No. 2816 has been unbolted and removed during the winter of 2005—2006 while repairs are made inside the Hudson locomotive.

816
15

Robert Ritchie chose the right man for the job of restoring No. 2816. Al Broadfoot had shepherded the renovation of numerous other locomotives: a shay, an English steam engine, and another locomotive on the White Pass and Yukon. In 1998 Broadfoot supervised movement of the neglected rusting hulk of CPR No. 2816 from Scranton back to Vancouver for her rebirth. At this time Broadfoot also maintained another Hudson, Royal Hudson, Class H1c, No. 2860 in Vancouver.

The designation of Hudsons in the CPR steam stable can be confusing. No. 2816 is considered a regular Hudson. Among the more obvious details of a regular Hudson is the absence of semi-streamlining effects, such as recessed headlights, smooth pilots, and hidden piping.

BELOW: The Canadian Pacific depot in Banff stands silent in the snow awaiting a possible return of CPR No. 2816 in the coming summer of 2006.

The Royal Hudson nomenclature came about in 1939 when King George IV and Queen Elizabeth visited Canada and the United States. Streamlined Hudson No. 2850 pulled their special train across North America. The railroad attached emblems of the royal crown to the front of the locomotive. The CPR liked this appearance and the historical connection to the British Empire. The railroad asked for, and received permission from, the King to place these emblems on all other Hudsons of this streamlined class. These streamed 4-6-4s forever became known as Royal Hudsons. Forty-five of these newer streamlined locomotives received this designation.

Although British Columbia Rail gave up passenger service and closed its steam shops in 2002, the West Coast Railway Association safely moved CPR No. 2860 to property. Broadfoot relocated east to Calgary to care for his latest charge, No. 2816 at CPR's Ogden Shops.

Above Right: The Banff Springs Hotel stands in glory against the backdrop of snowy mountains in the Canadian Rockies. The hotel continues to offer excellent service to travelers to Banff National Park throughout the year.

Above Left: The halls of the Hotel recall the glory days of Canadian Pacific passenger travel when the railroad constructed this magnificent hotel in the wilderness of the Canadian Rockies.

STEAM CHEST PRESSURE
BACK PRESSURE
WESTINGHOUSE
AIR BRAKE COMPANY
WESTLAND
MFG. CO.
RED HAND-MAIN RES.
WHEEL
SLIP
GEN
FLD
ALARM
NEUT
REV
EMERG
STOP

Left: The cab interior of CPR No. 2816, a regular Hudson class locomotive, undergoes shop repairs in the steam shops of the Canadian Pacific Railway in Calgary, Alberta, on January 20, 2006.

Above: Driving wheel and side rods of CPR Hudson No. 2816 stand still in the Calgary shops on January 30, 2006.

"CPR 2816 is used solely for public relations for Canadian Pacific," said Broadfoot, as he placed a wrench on the locomotive and thought about his career in steam railroading. "I've looked after the Royal Hudsons and Consolidation 3716, both CP engines, for 27 years." When asked if he had a specialty, Broadfoot replied, "I cover everything." Broadfoot ran the steam shops in Vancouver before coming to Calgary.

Broadfoot started in marine transportation in Vancouver. When the Hudsons came on the British Columbia Rail scene, Al couldn't stay away. At first his union in the marine industry prevented him from doing much besides sweeping the floors. But the siren song of steam railroading lured this Canadian mechanic away from the seas to the rails of the mountains.

In the Calgary shops today, Bill Stetler, now supervises the CPR steam program. Joining Stetler are other skilled craftsmen such as Jim Brogdan and Rene Clemens. The men seem intent on getting their "Lady" ready for the spring excursion season, which begins in May.

An air of professionalism pervades the CPR steam shop. And always looming above the workers is the grand machine, which can run at 75 miles per hour across the prairies and mountains of Canada.

2816

ABOVE: The Canadian Pacific logo graces the flank of CP No. 2816.

LEFT: With static test completed, a CPR worker walks around No. 2816 to the other side of the Hudson to continue his work. Calgary, Alberta, Ogden Yard, January 30, 2006.

Entering the steam shop at the massive Ogden Shops of the CPR in Calgary, the visitor is struck with the tidiness of the place. The jacketing is off the boiler and the faceplate has been unbolted and placed against the wall. But the locomotive appears clean, spic-and-span clean.

This Hudson is obviously the pride of the modern-day Canadian Pacific. Everyone speaks with enthusiasm when they describe the trips from the last two years with the locomotive called the "Empress." CPR Constable, Al Bittle, talks with Broadfoot about the trials and tribulations of the long trip over to Vancouver and back last summer, but it is obvious that Bittle enjoyed everyday of the journey.

Steaming past such Canadian Pacific icons as the Banff Springs Hotel and Chateau Lake Louise, both luxury lodges built in Banff National Park by the CPR in the early part of the twentieth century, the train provided rail photographers the chance for classic images from the past.

Through the winter of 2006, however, the crew of No. 2816 prepared the "Empress" for her next trip in the spring. This time the polished locomotive will steam east back to Montreal, the locomotive's birthplace. While snow lay deep in the Rocky Mountains to the West, the crew worked on boiler flues and hydraulic pumps.